Lecture Notes in Physics

Edited by H. Araki, Kyoto, J. Ehlers, München, K. Hepp, Zürich
R. Kippenhahn, München, D. Ruelle, Bures-sur-Yvette
H. A. Weidenmüller, Heidelberg, J. Wess, Karlsruhe and J. Zittartz, Köln

Managing Editor: W. Beiglböck

366

M.-J. Goupil J.-P. Zahn (Eds.)

Rotation and Mixing in Stellar Interiors

Proceedings of the Workshop
Frontiers in Stellar Structure Theory
Held in Honor of Professor Evry Schatzman
in Les Houches, France, June 19–25, 1989

Springer-Verlag
Berlin Heidelberg New York London
Paris Tokyo Hong Kong Barcelona

Editors

Marie-Jo Goupil
Observatoire de Paris, Section d' Astrophysique
5 Place J. Janssen, F-92195 Meudon Cedex, France

Jean-Paul Zahn
Observatoire Midi-Pyrénées, Université Paul Sabatier
14, Avenue Edouard Belin, F-31400 Toulouse, France

ISBN 3-540-53059-2 Springer-Verlag Berlin Heidelberg New York
ISBN 0-387-53059-2 Springer-Verlag New York Berlin Heidelberg

Printing: Druckhaus Beltz, Hemsbach/Bergstr.
Bookbinding: J. Schäffer GmbH & Co. KG., Grünstadt
2158/3140-543210 – Printed on acid-free paper

Preface

This volume constitutes the proceedings of a workshop held in honor of Professor Evry Schatzman, member of the Académie des Sciences, who was retiring in 1989 from his position at the Centre National de la Recherche Scientifique. The goal of the meeting is spelled out in its title "Frontiers of Stellar Structure Theory": it was to discuss recent progress made in some of the areas of stellar physics to which Evry Schatzman has contributed most in recent years: mixing in stellar interiors, redistribution and loss of angular momentum.

An informal committee (Annie Baglin, Jean Audouze and Jean-Paul Zahn) drafted the program in close collaboration with Evry Schatzman. They were joined by Marie-Jo Goupil, who took care of the local organization, and later of the proceedings. We chose les Houches as the meeting place because we knew that Evry Schatzman likes this wonderful setting very much, in the heart of the French Alps, where he has attended and organized several such workshops or summer schools.

Forty colleagues, including a few graduate students, responded to the invitation; others wrote how much they regretted not being able to come. Letters came from all over the world, expressing the high consideration of our community for Evry Schatzman, as a scientist and as a man.

We wish to thank all those who made this meeting possible: Professors Romestain and Boccara, the directors of the Ecole de Physique Théorique des Houches, for offering their hospitality; the local staff, and particularly Annie Giomot, for their patience, their kindness and their efficiency in solving the day-to-day problems; the Centre National de la Recherche Scientifique, for its generous financial support, through the research programs "Dynamique des Fluides Géophysiques et Astrophysiques" and "Structure Interne des Etoiles et des Planètes Géantes".

Finally, we express our warm thanks to the invited speakers and to all who, by their active and inspired participation, rendered this workshop so successful. We felt that their contributions should be shared by a larger number, and we are glad that Springer-Verlag agreed to publish these proceedings.

Paris and Toulouse
July 1990

Marie-Jo Goupil and Jean-Paul Zahn

PROFESSOR EVRY SCHATZMAN

Evry Schatzman: a new start

For us scientists, retirement has not necessarily the same implications as it has for many other people - who either fear it or long for it. During our entire professional life, we have the great privilege to organize ourselves as we wish, and when the moment comes to quit the "active" service, it often means that we can drop at last some teaching or administrative duties, and devote more time to the activity that we cherish most, namely research.

We know Evry Schatzman well enough to be convinced that it is just how he feels. For this reason we decided to hold a scientific meeting on the subjects which are the closest to his heart since several years, and to which he has enormously contributed himself.

He told us that it was Otto Struve who, in 1949, had attracted his attention to the rotation of stars: why do massive main-sequence stars spin so fast, whereas solar-type stars are such slow rotators? The answer was obviously linked to the loss of matter by the stars, as had been suggested by Fessenkov, but the required mass loss rates were incompatible with the observations. E. Schatzman understood the essentiel role of magnetic activity in this process: matter is channelled by the magnetic field to a considerable distance of the star (at what we call now the Alvèn radius), and it thus carries away much more angular momentum as if it were released at the surface. He announced his discovery in 1959; since then, he kept on refining the theory, by linking together rotation, magnetic field and turbulent convection.

He was also one of the first to recognize the importance of the anomalies in chemical composition at the surface of stars. He thoroughly explored a first explanation, that of nuclear reactions triggered by electromagnetic activity, and he rejected it when it became clear that the energy requirements could not be met. Then he started to examine in detail the separation processes due to gravity and the radiation field, and he came to the conclusion that these are so powerful that they must be inhibited by some other mechanism. We owe him the identification of this mechanism, and that is one of his outstanding contributions to the theory of stellar structure: according to him, it is turbulent mixing generated by the differential rotation of the stars.

A truly remarkable convergence between the two subjects in which Evry Schatzman has invested so much, it also emphasizes the importance of rotation in stellar evolution!

We believe that this meeting in les Houches has helped to clarify some key points of that exciting subject, and to initiate some further developments. Moreover, it was for us the opportunity to gather around Evry, to wish him all the best for this new phase of his brilliant and fecond carreer, and to express him our gratitude for what he has taught us and for what he has given to us.

L'invitation au Voyage

en l'honneur de Evry Schatzman

par Bérengère Dubrulle ....
avec la complicité de Charles Baudelaire

Astronome, mon frère
Songe à la douceur
D'aller là-haut sonder son cœur
Sonder à loisir
Pour y découvrir
Les mystères de ses heurts.
Le Soleil d'aujourd'hui
Pour ma théorie a les risques
De ces traîtres T Tauri
Brillantes à travers leur disque.

Là, tout n'est qu'ordre et beauté
Paradoxe, turbulence et mystères inviolés.

Des bancs de Lithium
Brassés par convection
Porteraient à l'Hélium
La botte de la déplétion
La dynamo toussive
La couronne poussive
Les supergranules, les plumes fractales,
Tout y miroiterait
Sous l'oeil discret
De ma lampe frontale.

Là, tout n'est qu'ordre et beauté
Paradoxe, turbulence et mystères inviolés.

VIII

Le champ magnétique
Affole ma boussole
Les neutrinos mystiques
Irradient mon piolet fol
Les éruptions solaires
Constellent ma lampe d'or
Les mystères solaires
Sont encore des mystères
Mais le monde s'endort
Dans sa chaude lumière.

Là, tout n'est qu'ordre et beauté
Attends un peu, Evry, avant de tout expliquer!

Contents

Introduction

The Observational Facts

Our Physical Knowledge

Interpreting the Observations

List of Participants

AUVERGNE Michel	Observatoire de Meudon 91195 Meudon Principal Cedex France
BERTHOMIEU Gabrielle	Observatoire de Nice BP 139 06003 Nice Cedex France
BOUVIER Jérome	Institut d'Astrophysique 98 bis, bld Arago 75014 Paris, France
CATALA Claude	Observatoire de Meudon 91195 Meudon Principal Cedex France
CAYREL Giusa	Observatoire de Meudon 91195 Meudon Principal Cedex France
CHARBONNEL Corinne	Observatoire Midi-Pyrénées 14, avenue E. Belin 31400 Toulouse France
D'ANTONA Francesca	Observatorio Astronomico di Roma I-00040 Monte Porzio Italy
DÄPPEN Werner	ESTEC Postbus 299 Noorwijk 2200A6 Netherland
DUBRULLE Bérengère	Observatoire Midi-Pyrénées 14, avenue E. Belin 31400 Toulouse France

GAVRYUSEVA Elena — Institute for Nuclear Research
60th October Anniversary Prospect Ave.
Moscou 117312
U.R.S.S.

GOUPIL Marie-Jo — Observatoire de Meudon
91195 Meudon Principal Cedex
France

HERBIG George — Institute for Astronomy
University of Hawaii
2680 Woodlawn Dr.
Honolulu, Hawaii 96822
U.S.A.

HOPFINGER Emil — Institut de Mécanique de Grenoble
B.P. 38402
Saint Martin d'Hères Cedex
France

KNOBLOCH Edgar — Department of Physics
University of California
Berkeley, CA 94720
U.S.A.

MAEDER André — Observatoire de Genève
CH-1290 Sauverny
Suisse

MANGENEY André — Observatoire de Meudon
91195 Meudon Principal Cedex
France

MASSAGUER Josef — Dept. Fisica Applicada
E.T.S.E. Telecomunicacio Apdo.
Correos 30.002
08080 Barcelona
Espagne

MAZZITELLI Italo — Instituto Astrofisica Spaziale
C. P. 67
I-00044 Frascati
Italy

MEGESSIER Claude	Observatoire de Meudon 91195 Meudon Principal Cedex France
MOFFATT Keith	University of Cambridge D.A.M.P.T. Silver street Cambridge CB3 9EW United Kingdom
PROVOST Jeanine	Observatoire de Nice BP 139, 06003 Nice Cedex France
RIEUTORD Michel	Observatoire Midi-Pyrénées 14, avenue E. Belin 31400 Toulouse France
SPRUIT Henk	Max-Planck Institut für Physik und Astrophysik Karl-Schwarzschildstr. 1 D-8046 Garching West Germany
SCHATZMAN Evry	Observatoire de Meudon 91195 Meudon Principal Cedex France
VAUCLAIR Sylvie	Observatoire Midi-Pyrénées 14, avenue E. Belin 31400 Toulouse France
VIGNERON Caroline	Observatoire de Meudon 91195 Meudon Principal Cedex France
ZAHN Jean-Paul	Observatoire Midi-Pyrénées 14, avenue E. Belin 31400 Toulouse France

INTRODUCTION

ROTATION, LITHIUM AND MIXING *

Evry Schatzman
Observatoire de Meudon
92195 Meudon Cedex, France.

Abstract. After a brief sketch of the lithium problem and a statement concerning the principles of its solution, the author presents the basic ideas : (1) the non-linear dynamo . The loss of magnetic energy is supposed to be due to buoyancy taking place in the neighborhood of the bottom of the convective zone. It is then assumed that the motion upwards of the flux tubes is taking place in a fluid having a high turbulent viscosity; (2) spin down. When carried into the equations describing the loss of angular momentum, with the assumption that the asymptotic value of the stellar wind is equal to the velocity of escape, it is found that the period of rotation increases like $(1 + t/t_0)^{3/4}$. Proofs of the law of spin-down are looked for :(i) in the differential rotation in the convective zone, (ii) in the statistical distribution of $v \sin i$ for F stars, and (iii) in the distribution of the periods of rotation as a function of mass in the Hyades. It is shown that the available sample of F stars is not homogeneous and is probably made of at least two sub-groups with a very different characteristic time t_0 of spin down. (3) pre-main sequence stars. It is suggested that electromagnetic braking begins during contraction along the Hayashi track when both a high dynamo number and a high magnetic Reynolds number characterize the properties of the outer layers. This leads to a relatively small velocity of rotation when reaching the main sequence, in contradiction with the observations in young clusters. (4) a consistent picture of lithium deficiency in late dwarfs is given, with lithium being carried to the level of nuclear processing by time dependent turbulent diffusion mixing. The high abundance of lithium in fast rotators in α Per is explained by the presence of the remnants of an accretion disk.

1. What about headache ?

The aim of this research has been, for many years, to build up a completely consistent theory connecting two different age effects : spin down of rotating stars and nuclear burning of light elements. During the last years, the amount of information concerning these two problems has considerably increased, and as well the number of theoretical papers.

---

* The original title was "Rotation and Mixing or how to eliminate ad hoc assumptions and provide headaches to astrophysicists ?"

The general aspect of the problem seems well clearcut : stars rotate more and more slowly with time (Schatzman, 1962; Kraft, 1967, 1969) and the surface abundance of lithium decreases: this is so well evident from the observations (Duncan and Jones, 1983; Butler et al, 1987; Boesgaard *et al*, 1988a; Balachandran *et al* , 1988; Boesgaard *et al*, 1988b; Cayrel *et al* 1984; Boesgaard and Tripico, 1986; Soderblom and Stauffer, 1984; Boesgaard, 1987; Hobbs and Pilachowski, 1986a; Hobbs and Pilachowski, 1986b; Spite *et al* , 1987; Garcia Lopez *et al* , 1988) that it is used as a measure of stellar age, in order to find the relation between age and rotation (Soderblom, 1983). However, there are shadows, fog and mist on this beautiful and simple landscape. What about the dip in lithium abundance found by Boesgaard and Tripico (1986) ? What about the high lithium surface abundance found by Balachandran *et al* (1988) in fast rotators in the young clusters α Per and the Pleiades ? What about the nuclear processing concerning lithium : only destruction or some production?

What about pre-main sequence lithium burning : the kee to lithium deficiency (Bodenheimer, 1965; D'Antona and Mazzitelli, 1984) or just a step in stellar history (Schatzman, 1984; Baglin *et al*, 1985) ? I shall try in the following to give the answer to some of these questions. The main idea is to provide a tool, a theoretical framework, giving, in principle, the possibility in the future of a better and refined approach of the problems, which will have to be considered then in greater details.

1.1. *The lithium problem.*

The main question which will be considered here concerns the surface abundance of lithium in the Sun and stars. It will be assumed that lithium burning is due to the transport of lithium by turbulent diffusion mixing at the level where lithium is nuclearly processed. Turbulence being due to differential rotation, we are led immediately to study rotation and its time dependence, in order to build the proper tool to study lithium burning.

After considering the physical problem of the source of macroscopic mixing in the radiative zone and its physical characteristics, we meet the mathematical problem of solving the equations.

It is well known that parabolic partial differential equations are more difficult to handle than ordinary differential equations or hyperbolic partial differential equations. This is perhaps the reason for which so many important publications on stellar structure and evolution ignore the physics of mixing. Mixing in convective zones is such an obvious mechanism taking place almost instantaneously at the stellar life time scale, that there is a great temptation to explain lithium anomalies by pre-main sequence burning : at least, this is the explanation which I can give to the attachment, for example in the paper of Balachandran *et al* (1988) to Bodenheimer's paper (1965). If the reader would allow me to be critical, it was quite natural, in 1965, to try to explain lithium anomalies by pre-main sequence evolution. It seems more difficult to understand it in 1989, when the large amount of informations on age effect on lithium abundances has now been collected and I even found surprising the last attempt of D'Antona and Mazzitelli (1984) to carry the same approach. On the contrary, the recent work of Proffitt and Michaud (1989) on the contribution of pre-main sequence burning to lithium abundance, without any attempt to explain everything all at once, seems to me to be an important and useful contribution. Let me make the hypothesis that the fear of macroscopic mixing is producing headache to many astrophysicists.

Consistency seems to me to be the most important thing, and I shall try, in the following, to show the way to consistency, even if it is difficult to be consistent in a field where there are still so many unknowns !

1.2. *Production or destruction ?*

This is a preliminary question. Does only lithium destruction take place, or is there any lithium production at the surface of stars? I rule out here, naturally, the production of lithium in red supergiants, which seems now well explained - at least in principle - by sudden heating during helium flash (Sackman *et al*, 1974) as I am concerned here only by main sequence and pre-main sequence stars. Spallation at the surface of stars, as the source of abundances anomalies, was in fashion in the early sixties, and has been extensively studied by Gradsztajn *et al* (1965) and Bernas *et al* (1967). Later on, Canal (1974) among others, has shown that, due to the very small value of the spallation cross sections, compared to the elastic cross section of fast protons, the explanation of abundance anomalies by spallation was raising difficult energetic problems. Considering the recent work of Kocharov and co-workers (1988) on the production of elements at the surface of stars by cosmic rays generated during solar-like flares, it is necessary to look again to the problem. I shall examin here only the order of magnitude of this effect.

The spallation cross section, as given by Foshina *et al* (1984), is

$$\sigma_i = \pi {r_0}^2 \, {A_i}^{2/3} \, (1 - T_P) \qquad ,$$

where r_0 is the proton radius, A_i is the mass number of the target nucleus i and T_P the nuclear transparency for protons. The actual cross section for the production of ^{7}Li from ^{12}C will be of the order of (1/6) of σ_i . In order of magnitude, it will be sufficient to take T_P equal to 0. With $r_0 = 1.1 \; 10^{-13}$ cm this gives $\sigma_i \cong 4. \; 10^{-27}\mathrm{cm}^{-2}$. The energy losses are given approximately by :

$$\frac{dE}{dx} \cong \frac{2\pi \, e^4 \, N_H}{m_e c^2} \ln\left(\frac{E}{kT}\right) \qquad ,$$

or

$$\frac{dE}{dx} \cong 10^{-29} N_H \qquad .$$

We shall assume that the energy loss is of the order of $E \cong 1$ GeV, and that the abundance ratio of Carbon to Hydrogen is of the order of 10^{-4}. This gives a total number of lithium nuclei produced by each proton of energy E:

$$N_{\mathrm{Li}} \cong \; 0.001 \; E \; \frac{N_{\mathrm{C}}}{N_{\mathrm{H}}} \qquad .$$

The total number of atoms of lithium present in the convective zone, with

$(N_{\rm Li}/N_{\rm H}) \cong 10^{-11}$ is of the order of 5. 10^{32}. The energy flux of cosmic rays at the surface of the Sun being :

$$\Im = \int_{E_{\rm min}}^{\infty} f(E)\, E\, dE \quad ,$$

where $f(E)\ dE$ is the energy spectrum of the cosmic rays, we can write for the total number of atoms of lithium produced in a time t :

$$N_{\rm Li} \cong 10^{-7}\ \Im\ t \quad ,$$

from which we derive the corresponding energy flux over 4.6 billions years:

$$\Im = \ 3.7\ 10^{11}\ \text{ergs cm}^{-2}\ \text{s}^{-1} \quad ,$$

which is about six times the actual heat flux at the surface of the Sun. We shall conclude that even the small number of lithium atoms presently visible at the surface of the Sun cannot have been produced by spallation . This is due to the efficiency of the stopping power of the solar plasma, to the small spallation cross section and to the very large size of the convective zone. The result would be very different if we had only to consider the solar atmosphere. With a mass per square centimeter 10^6 times smaller, this represent an energy flux small compared to the solar heat flux and this is compatible with the presence of lithium in the large flare of March 1989.

1.3. *Lithium as a function of time.*

This result brings us back to the problem of the time dependence of the abundance of lithium. Soderblom (1983) has well shown the correlation between age, rotation and lithium for late main sequence stars (from F to G spectral type) and the last results on lithium abundance in open clusters clearly confirm this general tendency (Baglin and Lebreton, 1990).

The problem is then to try to build a consistent picture, with the smallest possible number of parameters. This immediately raises several problems :

- what is the origin of the macroscopic mixing ? I shall assume in the following that the picture given by J.P. Zahn (1983a, 1983b, 1988) of the production of a 3-D turbulence is valid. The turbulent diffusion coefficient depends then on the period of rotation of the star. As there is obviously a process of spin-down, the consequence is that the turbulent diffusion coefficient is time dependent. Concerning the origin of the astrophysicist's headache (APHA), I think that one of the main problems today is to convince the scientific communauty that **there is** a turbulent flow which is induced by the rotation. Pinsonneault *et al* (1989) have taken into account the existence of this turbulent flow. However, they did not give a full description of the physics of the turbulent mixing.

- the spin down is related to the existence of a magnetic field at the surface of the star. What is the value of the magnetic field, and what is its dependence on the period of rotation and on the characteristic parameters of the star ? Answering these questions means developing the theory of a non linear stellar dynamo. Testing the dynamo theory is certainly of a great importance and is one of the ways of healing the headache even if the theory is exceedingly simplified.

2. The non linear dynamo.

The principle of the method is the following:

(1) assume a rate of growth of the magnetic field given by the linear theory;

(2) introduce a non-linear effect which will limit the growth of the magnetic field; in the present case, it will be considered that the limitation by buoyancy effect is the best assumption;

(3) the question now is to decide about the magnitude of the magnetic field and the size of the flux tubes on which the buoyancy acts. There is presently no way of determining these quantities from pure theoretical arguments. Some observations, the interpretation of solar magnetic activity, are guides which can help in making the proper choices. It is then necessary to justify them. I shall give a few simple arguments which have to be considered rather like anticipating on a theory to come than as real theoretical arguments.

(4) as we shall see, the astrophysical consequences, which will be presented in the next sections are in reasonable and even in good agreement with the observations. However, we shall keep in mind that, due to the very large number of parameters which are involved, it is not possible to state that nice astrophysical arguments are proving a physical theory, especially when it is so crude.

We start from the $\alpha-\omega$ dynamo equations in the spherical case. The vector potential **A** has only ϕ components, and the poloidal field B_P is given by :

$$\mathbf{B}_P = \ (\nabla \times A\ \mathbf{e}_\phi) \quad .$$

We have then for B_ϕ and A_ϕ (dropping the index ϕ) ,

$$\frac{\partial B}{\partial t} = s\ (\nabla\omega \times \nabla A)_\phi + s^{-1}\ \nabla\eta_T.\nabla s\ B + \eta_T\ (\nabla^2 - s^{-2})\ B \quad ,$$

$$\frac{\partial A}{\partial t} = \alpha B + \ \eta_T\ (\nabla^2 - s^{-2})\ A \quad ,$$

where $s = r\sin\theta$ is the distance to the axis of rotation , and α is the usual estimate (e.g. Moffatt, 1978, 1984),

$$\alpha \cong \frac{1}{3}\frac{\omega\ {l_T}^2}{H_\rho} \quad .$$

We are not so much interested by an exact solution of the rate of growth of the magnetic field. In order to obtain an order of magnitude of the rate of growth, we shall replace the derivatives with respect to θ and r respectively by iw and $i(q/R)$. Assuming $B \propto A \propto \exp(pt)$ we obtain,

$$pB = \quad iwA \frac{\partial \omega}{\partial r} + \frac{iq}{R^2} B \; \eta_T - \eta_T \; q^2 \; \frac{B}{R^2} - \eta_T \; \frac{B}{R^2} - \eta_T \; \frac{w^2 B}{R^2} \quad ,$$

$$pA = \; \alpha B + \; \eta_T \; \left(-\frac{q^2}{R^2} - \frac{w^2}{R^2} - \frac{1}{R^2} \right) \; A \quad .$$

The maximum of p is reached for $q = 0$,

$$w^2 = \; 2^{-(10/3)} \left(\alpha \frac{\partial \omega}{\partial r} \frac{R^4}{\eta_T{}^2} \right)^{2/3} \quad ,$$

and

$$p \cong \frac{\eta_T}{R^2} \; . \; 3 \; . \; 2^{-(10/3)} \; \left(\alpha \; \frac{\partial \omega}{\partial r} \; \frac{R^4}{\eta_T{}^2} \right)^{2/3} \quad .$$

We shall now consider the rate of losses. We shall write that the rate of loss of magnetic energy density is defined by buoyancy effects. We shall assume that the magnetic flux tubes have a tendency to move up and we shall consider the time scale of losses as defined by the time scale of buoyancy effect at the bottom of the convective zone. This problem has been considered for example by Schüssler (1977, 1979), after Parker (1975, 1977). However, they use the classical expression describing the motion of a cylinder in a low viscosity fluid. When the Reynolds number (av/ν) is large, the friction of the cylinder is due to the turbulent flow generated by the motion of the cylinder in the fluid, and the resistance per unit surface of the cylinder is proportional to ρv^2. We shall here also consider a cylindrical flux tube like a solid cylinder, but we shall suppose that, due to the turbulent flow in the convective zone, the viscosity of the fluid is very high. It is then possible to use the expression (Lamb, 1932) of the force acting on a cylinder moving in a viscous fluid at a velocity u . The rate of losses of magnetic energy is determined by the time scale of the ascending motion of the cylinder. The rate is of the order of (u/H) , where u is the velocity of the cylinder at the bottom of the convective zone and H is the vertical scale height. We have then for the rate of losses p_L :

$$p_L = \; \frac{u}{H} = g \; \frac{B^2}{8\pi P} \; \frac{\pi a^2}{4\pi\eta_T} \; \left(\frac{1}{2} - \gamma - \ln \left(\frac{1}{2} ka \right) \right) \; \frac{1}{H} \quad ,$$

where γ is the Euler constant, $(B^2/8\pi P)$ expresses the relative density deficiency $(\Delta\rho/\rho)$ in the flux tube, η_T is the turbulent viscosity, a the radius of the cylinder, and

$$k \; a = \left(\frac{u \; a}{\eta_T} \right) \quad ,$$

is supposed to be a small quantity.

We shall assume , following Schatzman and Ribes (1987) that, inside the dynamo active region, close to the bottom of the convective zone, the flux tube collapses into fibrils or ropes of radius a. Three arguments are in favour of a high magnetic field there. The first one is based on flux conservation and hydromagnetic equilibrium. Extrapolating from the surface of the Sun, where we see the sunspots, to the bottom of the convective zone, we must conserve the flux $B \; a^2$, the matter $\rho \; a^2$, and keep the pressure equilibrium. This provides a ratio (a/a_0) of the radius of the fibril inside the convective zone to the radius near the surface which is given approximately by

$$\frac{a}{a_0} \cong \left(\frac{\mu}{\Re \rho_0 T_0}\frac{B_0}{8\pi}\right)^{1/4}\left(\frac{\rho}{\rho_0}\right)^{-1/4} \quad ,$$

where $\Re$ is the gas constant. For a surface magnetic field of the order of 3000 Gauss, this gives at the bottom of the convective zone some thing of the order of 3.10^6 Gauss.

The second evidence comes from the analysis of the frequency splitting of the pressure modes by Dziembowski *et al* (1989) which gives, somewhere near the bottom of the convective zone magnetic fields of the order of a few million gauss in a very narrow region.

The third argument is a theoretical one. As summarized by Schüssler (1983), the magnetic field generated by the hydrogen convective zone dynamo is likely to be expelled just below the convective zone. This lead Schmitt and Schüssler (1989) to describe a dynamo confined to the bottom of the convective zone.

As we are concerned here by averages at a scale of several million years and not by shorter scales as the solar cycle, we shall finally write for the non linear dynamo,

$$2p = p_L \quad ,$$

the factor 2 coming from the fact that p concerns the rate of growth of the magnetic field and p_L the rate of losses of the magnetic energy. At this point, we just have to estimate properly the buoyancy effect.

We shall consider a local rate of losses due to buoyancy, which means that we assume that the magnetic field can find its way towards the surface of the star. The local rate of losses will be defined by the ratio of the upwards velocity due to buoyancy to a characteristic length, which we will ultimately take as H_P.

In order to find the characteristic time, we must take into account the fact that the dynamo theory gives an idea of the mean magnetic field, whereas we have to remember after looking to the solar properties that very likely the magnetic field divides itself into fibrils. The time scale of the buoyancy effect should be determined by the size of the fibrils and the strength of the magnetic field inside the fibrils.

We shall see, when writing the rate of loss of angular momentum, that we need to know the mean square value of the magnetic field at the surface of the star. On the other hand, we need also to know the relation between the magnetic field $< B^2 >$ as given by the dynamo model and the magnetic field inside the tube of force which experiences the buoyancy. We have shown that the magnetic field inside the fibrils, which we call B_f, gives a magnetic pressure comparable to the gas pressure outside the fibrils. We then assume flux conservation, so that the total flux at the surface of the star is equal to the total flux in the fibrils. The flux in the fibrils is of the order of $\pi\ R\ a\ B_f$; at the surface, half of the flux is an outgoing flux. We then have for the mean square value of the stellar surface field B^* :

$$4\pi^2 R^4 < B^{*2} > = 8\pi\ P_G\ \pi^2 R^2 a^2 \quad ,$$

or

$$a^2 = R^2 \frac{< B^{*2} >}{8\pi P_G} \quad ,$$

then, the rate of losses is :

$$p_L = g \frac{< B^{*2} >}{32\pi P_G} \frac{R^2}{H\eta_T} \left(\frac{1}{2} - \gamma - \ln\left(\frac{1}{2}ka\right)\right) \quad .$$

Replacing $(\partial\omega/\partial r)$ by $(\Delta\omega/H)$, where H is the scale of the convective zone, and $\Delta\omega$ the variation of ω over the convective zone we have:

$$\frac{1}{B^2}\frac{dB^2}{dt} = 2\left(\frac{1}{3}\frac{\omega\, {l_T}^2}{H_\rho}\Delta\omega\right)^{2/3} {\eta_T}^{-1/3} - C\, g \frac{< B^{*2} >}{32\pi P_G}\frac{r^2}{\eta_T}\frac{1}{H} \quad .$$

Growth, as well as losses, have a time scale much shorter than the evolution time scale. We can write $(dB^2/dt) = 0$ and we obtain then $< B^{*2} >$:

$$< B^{*2} > = \frac{64\,\pi P_G}{g}\frac{\eta_T}{r^2}\frac{H}{C}\left(\frac{1}{3}\frac{\omega}{H_\rho}{l_T}^2\,\Delta\omega\right)^{2/3} {\eta_T}^{-1/3} \quad .$$

We still need to obtain the values of $\Delta\omega$ and of C. We shall write that the Coriolis force in the convective zone is balanced by the viscous shear flow associated with $\Delta\omega$:

$$2 < \mathbf{u}_T \times \omega > = \frac{\eta_T\,\Delta\omega}{H} \quad .$$

If we consider that the supergranulation is providing the largest contribution to the Coriolis force, we obtain :

$$< \mathbf{u}_T \times \omega > = \frac{1}{6}\frac{d^2}{R^2} u_T\,\omega \quad ,$$

where d is the radius of the supergranulation cells. We then obtain the equation for ka and can derive the value of C. We can consider that we have the model of the non-linear dynamo.

We have now to relate the value of B^* to the rate of loss of angular momentum, to the lithium burning and to show that the observations can be explained quantitatively by this model.

3. Loss of angular momentum.

The loss of angular momentum depends on the rate of mass loss and on the Alfvenic distance, distance at which the loss of angular momentum takes place. It depends also on the structure of the surface magnetic field (dipole, quadrupole, or something else), as shown by Roxburgh (1983) and by Mestel and Spruit (1987). In the solar case, with a magnetic fields of a few thousand gauss in sunspots, the magnetic field of the system of dipolar sunspots is larger than the magnetic field of the solar dipole. The angular dependence of the dynamo can be estimated by considering the quantity w , which is :

$$\mathrm{w} = 2^{-5/3} \left(\frac{1}{9} \frac{\omega^2 {l_T}^2 d^2}{H_\rho} \frac{u_T R^2}{{\eta_T}^3} \right)^{1/3} .$$

With the solar values, we obtain w $\cong 1.7$ which is quite close to the actual poloidal scale, which corresponds to w $\cong 2$. The meridional structure for faster rotators, which decreases like $\omega^{-(2/3)}$, is obviously smaller. In such conditions, it seems difficult to establish clearly the power of ω which is involved, according to Roxburgh (1983) and to Mestel and Spruit (1987), in the process of angular momentum loss. This is the reason for which I shall prefer the formulation of Durney and Latour (1978) , who assume that, in the stellar wind, the magnetic field decreases like r^{-2} and that the asymptotic velocity of the stellar wind equals the escape velocity from the star. It has the advantage that it eliminates the rate of mass loss, and is written :

$$\frac{d}{dt} KMR^2\omega = -\frac{2}{3} (B^* R^2)^2 \left(\frac{GM}{R} \right)^{-1/2} \omega ,$$

where KMR^2 is the angular momentum. Assuming solid body rotation, we derive the expression for the rate of decrease of the angular velocity :

$$\frac{1}{\omega^{7/3}} \frac{d\omega}{dt} = -\frac{1}{KM} \frac{128\pi}{3} \frac{1}{g} P_G \frac{H}{C} \left(\frac{GM}{R} \right)^{-1/2} \left(\frac{1}{9} \frac{d^2}{R^2} \right)^{2/3} ({l_T}^2 u_T)^{2/3} .$$

It is clear that the right handside is only model dependent, and for main sequence stars depends mainly on the mass and is a little time dependent through the change of luminosity and depth of the convective zone.

The logarithmic dependence of C upon (ka) gives an easy possibility of estimating its value .

3.1 *Size of the super granulation.*

In order to check the validity of this formulae, we shall first consider the Sun and shall derive, from the present rate of loss of angular momentum, the size of the cells of the supergranulation. From standard description of the convective zone we derive the following estimate of the turbulent velocity near the bottom of the convective zone :

$$F_{\mathrm{conv}} = F_{\mathrm{total}} \cong 20 \rho u_T^3 .$$

With $\omega \cong 3.\ 10^6$, $(1/\omega)(d\omega/dt) \cong (10^9 \mathrm{years})^{-1}$, $C = 25$, $K = (1/14)$ we obtain

$$\mathrm{d} \cong 7000 \text{ km} ,$$

which is a reasonable and unexpectedly correct value.

3.2 *Periods of rotation as an age effect.*

Let us write the rate of loss of angular momentum :

$$\frac{1}{\omega^{7/3}} \frac{d\omega}{dt} = -f(M) \quad .$$

The solution can be written :

$$\omega = \omega_0 \left(1 + \frac{4}{3} f(M) \, {\omega_0}^{4/3} \, t\right)^{-3/4} \quad ,$$

or

$$\omega = \omega_0 \left(1 + \frac{t}{t_0}\right)^{-3/4} \quad ,$$

with

$$t_0^{-1} = \frac{4}{3} f(M) \, {\omega_0}^{4/3} \quad .$$

The question now is naturally to check the $t^{-3/4}$ law, which differs from the Skumanich relation (1972). Bohugas *et al* (1986) have derived from the observational data a $t^{-3/4}$ law, and it is possible to fit the same law to the data of Simon *et al* (1984) and of Benz *et al* (1984). However all these methods depend on the age determination of the stars, and even the age determination of open clusters is questionnable, as it depends on their absolute magnitude. The age of the Pleiades has been recently multiplied by 2 (Mazzei and Pizatto, 1989).

The data collected by Boesgaard and Tripico (1986) on field stars opens another possibility, as they can be interpreted as the combination of the distribution of the direction of the axis of rotation and of the age distribution with its influence on the velocity of rotation through the spin down effect. Let us assume that all stars in a small spectral interval start their life with the same initial equatorial velocity v_0.

What is observed is a certain velocity v which can be written :

$$v = v_0 \sin i \left(1 + \frac{t}{t_0}\right)^{-3/4} \quad .$$

The number of stars in the velocity interval dv is proportional to the time spent with a given velocity . We then have the number of stars in the interval $di \, dv$:

$$d^2 N = \frac{1}{3\left(\left(1 + t_1/t_0\right)^{1/4} - 1\right)} \frac{{v_0}^{1/3} \, dv}{v^{4/3}} \sin^{4/3} \, i \, di \quad ,$$

where t_1 is the maximum time spent on the main sequence in a small spectral interval. It is necessary to integrate $d^2 N$ with the condition of a given value of v, which gives :

$$dN = \frac{1}{3} \frac{{v_0}^{1/3} dv}{v^{4/3}} \int_{\sin i = \frac{v}{v_0}}^{\sin i = \min(Q\frac{v}{v_0}, 1)} \sin^{4/3} \, i \, di \quad ,$$

where $Q = (1 + (t_1/t_0))^{3/4}$ is related to the final equatorial velocity v_1 after the time t_1 spent on the main sequence, $v_1 = v_0\ Q^{-3/4}$.

The cumulative distribution function $F(v)$ is obtained by integration from v = 0 to v = v. Obviously, for v small, F is proportional to v^2:

$$F = \frac{1}{14}\frac{v^2}{{v_0}^2}\left(\left(1+\frac{t_1}{t_0}\right)^{7/4} - 1\right) \quad \text{for} \left(\frac{v}{v_0}\right) << 1 \quad .$$

Let us compare the cumulative distribution function taken from the data of Boesgaard and Tripico (1986) to the theoretical one (fig. 1). Assuming that the characteristic time t_0 is equal to 2.36 Gyr for a life time of 3.3 . Gyr on the main sequence, this gives $Q =\ 1.5$. With an initial equatorial velocity of 40 km s^{-1} this corresponds to a final equatorial velocity of about 27 km s^{-1}. With these numbers, and for small values of (v/v_0),

$$F \cong\ 0.78\ (v/v_0)^2 \quad .$$

This theoretical curve is definitely below the observed cumulative distribution function. Measurements of the equatorial velocity has an observational bias, the overestimate of the equatorial velocity. This means that after correction, the number of slow rotators would be even larger, increasing the discrepancy with the theoretical curve. It appears definitely that there is a large excess of slow rotators among the set of F stars studied by Boesgaard and Tripico (1986).

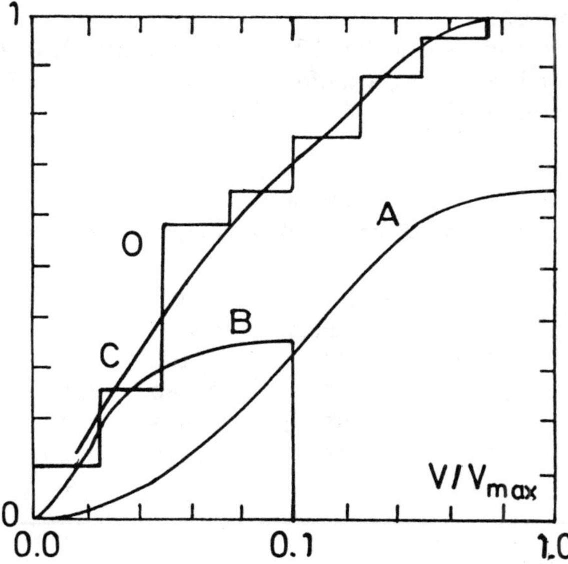

Figure 1. The integrated histogram of v sin i for early F dwarfs (Boesgaard and Tripico, 1986). Curve O represents the observations. Curves A and B represent the two populations (F0 to F4, and F5-F6) and curve C is the sum of curves A and B.

Without the time dependence of the equatorial velocity, the situation is even worse. It is suggested here that the sample of Boesgaard and Tripico is in fact made of two populations which we shall call A and B. Population A (fig. 1) is typical of F stars, with a slow rate of loss of angular momentum; it is assumed here that it represents 65% of the sample. Population B (fig. 1) is made of stars reaching the main sequence with a lower velocity (e.g. 20 km s^{-1}) and having a much larger rate of loss of angular momentum, with $Q = 4$. This represents a ratio of the life time on the main sequence to the time scale of angular momentum loss $t_1/t_0 = 5.34$. This value corresponds to late F type stars. It is assumed here that it represents 35% of the sample in agreement with the list of Boesgaard and Tripico. Curve C (fig. 1) is the addition of curve A and B and seems to fit the observed cumulative distribution function (fig. 1).

3.3 *Periods of rotation as a function of mass.*

This is a test of the quality of the function $f(M)$ of the mass which determines the time scale t_0 of the rate of loss of the angular momentum. In order to carry the scaling of the function $f(M)$ we assume a principle of self-similarity , and write :

$$f(M) \approx \frac{R^{5/2}}{M^{5/2}} \left(\frac{r}{R}\right)^2 P_G \, {H_P}^{7/3} \left(\frac{R-r}{R}\right)^{4/3} \frac{L^{2/9}}{r^{4/9} \rho^{2/9}} \; .$$

We have checked that $f(M)$ gives a correct solar value. For the low mass stars, there is an asymptotic value of the period of rotation which is independent of the initial value. This approximation is not valid when the time scale of the spin down is comparable to the life time of the stars.

With the stellar envelopes calculated by A. Baglin (1988), it is possible to obtain the values of $f(M)$ and, assuming a uniform initial period $P_0 = 3$ days, the following distribution of the actual periods (table 1 and fig. 2) .

Table 1

Periods of rotation and spin down

M	T_{eff}	L	f/f(0.9)	P	t_0
solar units		solar units		days	10^8 years
0.9	5140	0.37	1	11	2.3
1	5530	0.62	0.766	9.37	3
1.1	5870	0.95	0.513	7.49	4.5
1.2	6160	1.48	0.155	4.50	14
				3*	

* asymptotic value

We shall consider at this point that the expression of the function $f(M)$ can be considered as a reasonable basis for studying the evolution of rotating stars.

4. Spin down of pre-main sequence stars.

We shall follow here the presentation of Schatzman (1989).

Pre-main sequence stars are fully convective and the scaling based on the size of the convective zone is not any more possible . In order to estimate the characteristic magnetic field, we shall consider the depth at which the dynamo is the most efficient. In an entirely convective star, we can expect a structure in depth of the magnetic field.

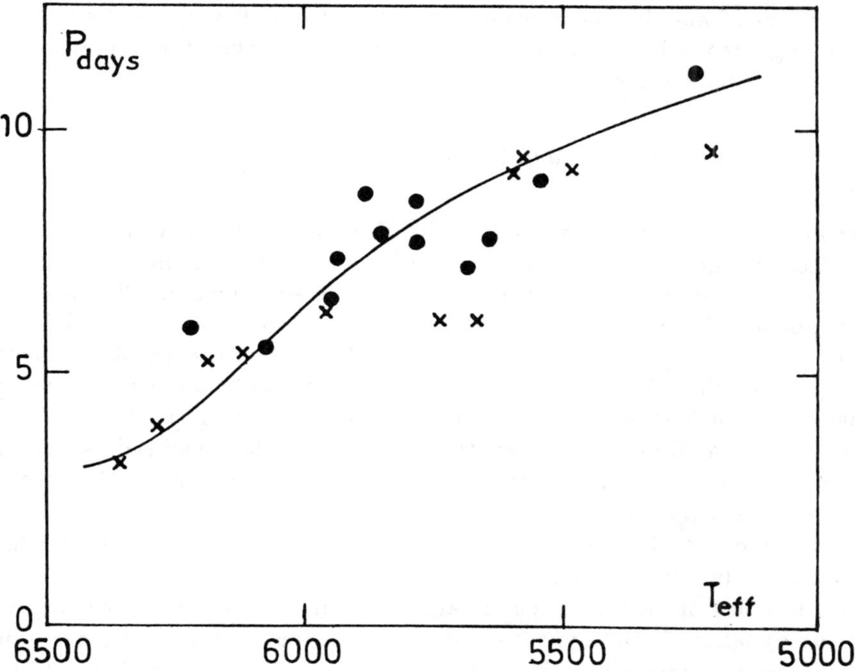

Figure 2. The periods of rotation as a function of mass in the Hyades. The observational points are from Rebolo *et al* (1988). Black circles correspond to observed periods.

This can be related to the scaling of the $(\alpha\omega)$ dynamo term,

$$\frac{1}{12}\ \omega\ H_{\rm turb}\ \frac{d{\rm ln}\omega}{dt}\ \frac{r\ \omega}{\eta_T}\ \frac{r^4}{{\eta_T}^2} \qquad .$$

In the outer part of the star, the scale height H_{turb} behaves like r^2T, the temperature is proportional to $R - r$, and the derivative with respect to r of the $(\alpha\omega)$ term has a maximum at $r = 0.8\ R$. The corresponding pressure scale height is $H_P = 0.08\ R$. Introducing these values in the expressions for the increased rate of the magnetic field, it is possible to obtain an order of magnitude :

$$< B^2 >= 71.6\ C\ \frac{H_{\mathrm{turb}}}{g} \left(\frac{l_T}{g}\ \omega\ \frac{d\omega}{dt} \right)^{2/3} P_G \quad .$$

We can calculate the constant C of the buoyancy, assuming a period of 4 days and a time scale of the spin down of 0.5 Gyr. We obtain for the m.s.r. of the surface magnetic field 230 $R^{*1/2}$ gauss, which is in agreement with the values given by Shu *et al* (1988) in their model of T Tau stars. If we carry these expressions into the equation which gives the rate of loss of angular momentum, we obtain the following solution. Introducing the initial angular velocity and the radius R_i of the star when the dynamo turns on we obtain for a one solar mass star :

$$\omega = \omega_i \left(1 + \ 0.653 \left(1 - \left(\frac{R_i}{R_0} \right)^{-1.951} \right) \right)^{-1/4} \quad ,$$

ω is the angular velocity at the end of the Hayashi track. Beyond that point, it is necessary to take into account the evolution of the momentum of inertia, which becomes smaller, when the star becomes mainly in radiative equilibrium. We have assumed a time dependence of the radius $r \approx t^{-0.155}$, as derived from the tables of D'Antona and Mazzitelli (1984). As we can see, there is a slow decrease of the angular velocity during the contraction, which in fact means a considerable loss of angular momentum. For an initial radius of 8 solar radii, an initial angular velocity given by the condition of rotational breaking, we obtain, when the star reaches the solar radius, an equatorial velocity of 17 km s^{-1}. All these numbers are reasonable, and show that with this model we are on the right track.

The question, at this point, is to determine the radius at which, during the contraction, the dynamo turns on.

From the point of view of a logical argument, it is absolutely necessary to have a start of the dynamo. If the dynamo had been working from the very early beginning of the pre-main sequence phase, it would be difficult to explain stellar rotation. On the other hand, an accretion disk like those present in T Tau stars (see for example the paper of Bertout *et al*, 1988) could provide angular momentum and compensate the losses : this is the point of view of Shu *et al* (1988). However, all pre-main sequence stars do not have accretion disk.

I have suggested (Schatzman, 1989) that the appearance of the magnetic field at the surface of the star and, consequently, the beginning of electromagnetic braking, is possible if two conditions are fulfilled :

(i) the dynamo number should be larger than 1

(ii) the magnetic Reynolds number should be larger than 30 (J. Leorat, 1975).

This last condition means that the electric conductivity is such that the electric currents generated by the dynamo process are not dissipated by Joule effect.

A critical property of stellar matter has to be taken into account : above $10^4 K$, the matter is fully ionized and is a good conductor; below, the free electrons are provided only by the metals and the conductivity of the matter is 10^5 smaller.

I suggest that these physical conditions are satisfied when the level on which the dynamo condition is fulfilled is also the level where the condition on the magnetic Reynolds number becomes valid, at a temperature of $10000 K$. This can be turned into a condition on the equatorial velocity. The dynamo mechanism becomes efficient when

$$v_{\mathrm{equ}} > 125 \ \mathrm{km\ s^{-1}}\ L^{*-1/4}\ M^{*7/3} \quad .$$

If the velocity is smaller, the region where the dynamo number is larger than one, $N_{\mathrm{dynamo}} > 1$, falls into the region where the magnetic Reynolds number is small. Therefore, the dynamo mechanism cannot be operating near the stellar surface, no magnetic field will appear at the surface, and no electromagnetic braking will take place. Conversely, if the equatorial velocity is large enough, the dynamo turns on and the electromagnetic braking starts.

5. Lithium burning.

In some sense, all what is written above has only one purpose : explaining the lithium deficiency in the low main sequence stars. What we have to take into account here, is that the turbulent diffusion coefficient is time and space dependent, and that the local rate of lithium burning is space dependent. The general indications which can be obtained from a constant diffusion coefficient, and with the use of simplified boundary conditions are not valid, and moreover, it has been shown by Baglin and Lebreton (1990) that they do not fit with the observations. The main problem is to fit the mass dependence of lithium burning with the observations.

5.1. *An analytical approach.*

In principle, only a numerical solution of the diffusion equations could provide the final answer to this question. However, it is worth to try an approximate analytical solution. It will be, anyhow , an indication of validity of the model.

The expression of the turbulent diffusion coefficient given by Zahn (1988) will be used, neglecting the possible effect of the inversion of the direction of the circulation, which - anyhow - would take place in the convective zone and therefore would not have any influence on the mixing process. We write :

$$D_T = \frac{1}{5} \frac{L\,\omega^2\, r^6}{G^2\, M^3} \frac{1}{\Delta\nabla} \quad . \tag{5.1}$$

Should we take into account the reversal of the circulation at the level where the quantity $(\omega^2/2\pi G\rho)$ becomes equal to one ? We ignore the factor

$$\left|\left(1 - \frac{\omega^2}{2\pi G\rho}\right)\right| \quad ,$$

which is of importance only if the convective zone is very shallow. The validity of this factor, according to a remark of Zahn (1989) deserves some discussion. Anyhow, we shall assume a time dependence of ω, resulting from the theory of the spin-down, $\omega \propto (1+(t/t_0))^{-3/4}$. We shall show that a solution of the diffusion equation, inspired by the WKB method is a reasonable approximation.

Let us recall before that the simplified method which has been used by Schatzman (1981) and by Schatzman and Maeder (1981) consists in assuming that at a certain level the lithium concentration vanishes. With a distance h from the bottom of the convective zone to the lithium burning level, a turbulent diffusion coefficient D_T, and a concentration decreasing exponentially, $c = c_0 \exp(-pt)$, p is a solution of the eigenvalue equation :

$$\sqrt{\frac{p\,h^2}{D_T}}\ \mathrm{tg}\ \sqrt{\frac{p\,h^2}{D_T}} = \frac{h}{H_P} \quad . \tag{5.2}$$

Whereas it can be shown that equation (5.2) gives a satisfactory order of magnitude for p , this model is obviously not valid when the bottom of the convective zone becomes close to the lithium burning level. Equation (5.2) is derived from a too much simplified model to give properly the behaviour of p as a function of mass. The temperature dependence of the rate of lithium burning around 2. 10^6K, $dc/dt \propto T^N$ with $N \cong 25$ means that the concentration does not vanish at a well defined level. The diffusion equation must include the term of chemical reactions :

$$D_T\,\frac{\partial^2 c}{\partial z^2} = \frac{\partial c}{\partial t} + K(z)\,c \quad . \tag{5.3}$$

In principle, only numerical solutions are significant. However, equations of this kind are frequently met in physics, and are usually solved analytically by the WKB approximation. This implies that the vertical scale heigh $\partial z/\partial \ln c$ is small compared to the scale of $D_T(z)$ and $K(z)$. However, even when this quantity is not fullfilled, the WKB approximation is not far from the exact solution.

The next question concerns the time dependence of D_T. With the expression (5.1) the ω^2 term gives a dependence in $(1+(t/t_0))^{-3/2}$. We must also notice that we have an r dependence almost exactly in r^6 in the small interval from the bottom of the convective zone to the burning level of lithium. $\Delta\nabla$ vanishes at the boundary. This is not the source of any singularity. It grows quickly to a value of the order of 0.15 which we shall use. In the integral $\int {D_T}^{1/2}\,dr$, the contribution of the region where $\Delta\nabla$ is small is negligible.

Coming back to the diffusion equation which we write more explicitly in spherical coordinates :

$$\frac{1}{r^2}\frac{\partial}{\partial r}\left(r^2 D_T\,\rho\,\frac{\partial c}{\partial r}\right) = \rho\,\frac{\partial c}{\partial t} + K(r)\,\rho\,c \quad , \tag{5.4}$$

where $K(r)$ includes the ρX factor, X being the abundance of hydrogen. With an r^6 law for D_T, the method consists in trying a solution of the form $c = r^n \phi$.

The equation becomes :

$$\frac{\partial^2 \phi}{r^2} + \left(\frac{2n+8}{r} + \frac{d\ln\rho}{dr}\right)\frac{\partial\phi}{\partial r} + \frac{1}{r}\left(\frac{n(n+1)}{r} + n\frac{d\ln\rho}{dr}\right)\phi = \frac{1}{D_T}\frac{\partial\phi}{\partial t} + \frac{K}{D_T}\,\phi \quad . \tag{5.5}$$

With the choice $n = -4 - (1/2)(d\ln\rho/d\ln r)$ it turns out that the $(1/r^2)$ term is then negligible. Proceeding with a WKB solution of the form $\exp(-pt)$, where p is supposed to be a slowly varying function of t, the ϕ equation reduces to

$$\frac{\partial^2 \phi}{\partial r^2} + \frac{p - K(r)}{D_T} \phi = 0 \quad . \tag{5.6}$$

The approximate WKB solution for $p > K$ is

$$\phi = \left(\frac{D_T}{p-K}\right)^{1/4} (A \cos \Phi + B \sin \Phi) \quad , \tag{5.7}$$

with

$$\Phi = \int_r^{r_1} \sqrt{\frac{p-K}{D_T}}\, dr \quad , \tag{5.8}$$

where r_1 is the lower boundary of the convective zone.

For $p < K$,

$$\phi = \left(\frac{D_T}{K-p}\right)^{1/4} A e^{-\Psi} \quad , \tag{5.9}$$

with

$$\Psi = \int_r^{r_2} \sqrt{\frac{K-p}{D_T}}\, dr \quad , \tag{5.10}$$

where r_2 is the radius where $K - p = 0$

The boundary conditions are :

-at the bottom of the convective zone, there is continuity in the flux of the test element. If we assume that the density law in the convective zone can be written $\rho \propto (R-r)^{\alpha}$ we have :

$$D_T \frac{\partial c}{\partial r} = pH_\rho\, c + \frac{\alpha+1}{N+\alpha+1}\, p_1\, H_\rho\, c \quad , \tag{5.11}$$

where H_ρ is a measure of the mass per unit surface in the convective zone. The second term on the right represents the rate of lithium burning in the convective zone; p_1 is the rate of burning at the bottom of the convective zone. In practice $\alpha = 1.5$, $N = 25$ which for the coefficient in the right hand side gives the value (1/11).

- at the boundary $(p - K) = 0$, there is a continuity of c and its derivative :

$$\left(\frac{1}{c}\frac{\partial c}{\partial r}\right)_{r_{2+}} = \left(\frac{1}{c}\frac{\partial c}{\partial r}\right)_{r_{2-}} \quad . \tag{5.12}$$

We have to fit the solutions (5.7) and (5.9) at the boundary $r = r_2$. Introducing the notations :

$$\begin{cases} a = \dfrac{p_1 H_P^2}{D_T} \\ x = \dfrac{p H_P^2}{D_T} \end{cases} \tag{5.13}$$

and taking into account Jeffrey's connection formula (Heading, 1962), we obtain the relation:

$$tg\left(\Phi_2 + \frac{\pi}{4}\right) = \frac{x - \sqrt{(x-a)} + \left(\frac{H_P}{r_1} - \frac{a}{11}\right)}{-x - \sqrt{(x-a)} - \left(\frac{H_P}{r_1} - \frac{a}{11}\right)} \quad . \tag{5.14}$$

An approximate expression of Φ_2 is :

$$\Phi_2 = \int_{r_2}^{r_1} \sqrt{\frac{p-K}{D_T}}\, dr \cong \frac{z_1}{H_P}\sqrt{x}\left[\left(\frac{x}{a}\right)^{1/N} - 1\right] \quad , \tag{5.15}$$

z_1 is defined in the following way. Assuming a linear relation between the temperature and the distance z to a reference level , which may be different from the surface of the star , we obtain the relation :

$$\frac{z_1}{H_P} = \left(\frac{\Gamma}{\Gamma - 1}\right)_{\rm Rad} \cong 4 \quad .$$

As a turns out to be small, or small compared to x, the relation between a and x can be written :

$$a = x\left[1 + \frac{1}{4\sqrt{x}}\left(-\frac{\pi}{4} + \arctan\frac{x - \sqrt{x} + q}{-x - \sqrt{x} - q}\right)\right]^{-N} \quad , \tag{5.16}$$

with

$$q = \left(\frac{H_P}{r_1} - \frac{a}{11}\right) \quad .$$

The relation (5.16) shows little sensitivity to the value of q, and this gives the possibility of ignoring the effect of the presence of a in the expression of q.

Let us now consider the numerical values. The first point concerns the value of the quantity H_P, which in fact is a measure of the amount of lithium which is stored in the convective zone. The approximation of the plane parallel convective zone is not valid, and the real value of H_P is not $\Re T/g\mu$. For $\gamma = 5/3$, and a radius r_1 of the boundary of the convective zone, we have :

$$H = 2\,(R - r_1)\left[\frac{1}{5} + \frac{2}{35}\,\frac{R - r_1}{r_1} + \frac{8}{315}\left(\frac{R - r_1}{r_1}\right)^2\right] \quad ,$$

which, in the solar case gives for an adiabatic convective zone :

$$H = 0.454\,(R - r_1) \quad ,$$

which has to be compared to the relation between the pressure scale height and the thickness of an adiabatic surface convective zone :

$$H_P = \frac{4}{10}\,\frac{r_1}{R}\,(R - r_1) \quad .$$

With the relation (5.13), changing H_P into the new H, using an interpolation formula

$$x = X \left(\frac{a}{A}\right)^m \quad ,$$

we obtain :

$$\int_0^{t_1} p\, dt = t_0 \left(\frac{p_1}{A}\right)^m X\, H^{-2(1-m)} {D_1}^{1-m} \frac{2}{(1-3m)} \left[1 - \left(1 + \frac{t}{t_0}\right)^{-\frac{1-3m}{2}}\right] . \quad (5.17)$$

It is either possible to introduce in this expression the present value D_1 of the turbulent diffusion coefficient in the Sun and to derive the actual depletion factor, or to derive from the actual value of the depletion factor $e^{-4.6}$ the turbulent diffusion coefficient and then to derive for example the initial period of rotation of the Sun. If we choose the second method, we have :

$$\begin{aligned} m &= 0.260 \quad , \\ X &= 4;\ A = 0.014088 \quad , \\ t_0 &= 0.3 \text{ Gyr} = 9.45\ 10^{15} \text{ s};\ t_1 = 4.6 \text{Gyr} \quad , \\ T_1 &= 1.95\ 10^6 \text{ K};\ \rho_1 = 0.18 \text{ g cm}^{-3} \quad , \end{aligned}$$

and with

$$p_1 = \rho_1\, X_H \frac{8.04\ 10^8}{{T_9}^{2/3}} \exp\left(-\frac{8.471}{{T_9}^{1/3}}\right) \quad ,$$

or

$$p_1 = 2.318\ 10^{-20} \quad ,$$

we find then $D_1 = 490$. With the expression of Zahn of the turbulent diffusion coefficient, we obtain $P_1 = 26.6$ days, and finally the initial period of rotation $P_0 = 3.27$ days. It must be well kept in mind that without a detailed numerical computation, it is not possible to affirm that there is a perfect fit between the theory and the observations. However, it is clear that the solar value of the lithium depletion puts a strong constraint on the depth of the convective zone, which determines both p_1 and H. In the expression (5.17) the integral $\int p\, dt = s$ varies like ${z_1}^{5.02}$ for $m = 0.260$. A variation of 10% on s or on other factors means a variation of z_1 of 2%, which gives an idea of the precision which is needed in the definition of the depth of the fully mixed region.

5.2 *The mass dependence.*

Let us consider the mass dependence of the exponent in the exponential which gives the deficiency, $(c/c_0) = \exp(-s)$. It seems at first look on equation (5.17) that the most important effect is the growth of the depth of the convective zone, which provides an increase of the temperature at the bottom of the convective zone for decreasing stellar masses.

With the tables of the outer layers computed by A. Baglin (1988), it is possible to calculate $s(M^* = 1) = 2.288$. With different values of the parameter X, A, and m

($X = 10$, $A = 0.3172$ and $m = 0.393$), one finds $s(M^* = 0.9) = 4.685$. In this interval, s varies like $M^{*-6.80}$. With a variation of the effective temperature $T_{\rm eff} \propto M^{*0.60}$ this gives for $T_{\rm eff} = 5560$ K ($M^* = 1$) and $T_{\rm eff} = 5000$ K respectively,

$$s\,(5560K) = 2.29 \quad ,$$
$$s\,(5000K) = 7.63 \quad .$$

The corresponding depletions are respectively 0.101 and 0.000486 instead of 0.1 and 0.001 (Cayrel *et al*, 1984; Boesgaard and Tripico, 1986). This can be considered as a remarkable agreement, if we remember the extreme temperature sensitivity of p_1, the rate of destruction of lithium at the bottom of the convective zone and the uncertainties of the internal structure models.

5.3. *Young clusters anomalies.*

The results of Stauffer *et al* (1988) and of Balachandran *et al* (1988) about the young cluster α Per raise many questions. The presence, among fast rotators in α Per, of lithium rich stars (Balachandran *et al*, 1988), the abundance being either cosmic or slightly deficient, is quite surprising. With an equatorial velocity as high as 100 times the solar one, the turbulent diffusion coefficient should be 10 000 larger than the solar one. The time scale of lithium burning would be of the order of 10^5 years, hundred times less than the age of the cluster. How can we explain the presence of lithium with the cosmic abundance (or half of it) inside fast rotators ?

We can consider several possibilities:

(i) only the surface is rotating fast and the slow rotation inside is compatible with a slow depletion of lithium; however, this is contradictory with the fact that the slow rotators in α Per are strongly deficient in lithium;

(ii) the turbulent diffusion coefficient vanishes in the radiative zone, at the level **C** where $\omega^2 = 2\pi G\rho$. We have to consider here two questions : (1) does that level sit below the bottom of the convective zone ? (2) the the radial component of the meridional circulation vanishes at the same level **C** but the latitudinal component does not vanish. Does the turbulent diffusion coefficient P_T actually vanish ? The second derivative $\partial^2 v_\theta/\partial r^2$ vanishes at the level **C** and this is a well known condition of stability of the shear flow : this is the answer to the first question. The reversal of the meridional circulation takes place below the convective zone only for early spectral types. For a mass $M^* \cong 0.9$, $T_{\rm eff} = 5140$ K, this would imply $v_{\rm equ} > 230$ km s^{-1}. This means that for later spectral types, say for $T_{\rm eff} < 5200$ K it is necessary to find another explanation (Table 2).

(iii) the hypothesis of the presence of a disk (Schatzman, 1990) explains both the high equatorial velocity and the presence of lithium. The accreting disk provides both angular momentum and lithium. There is a balance between the loss of angular momentum due to electromagnetic breaking and the capture of angular momentum from the disk. The abundance of lithium is then the result of a balance between the lithium brought by accretion and the fast destruction inside the star. If p is the rate of lithium destruction, X_7 the concentration in the disk, and $X_7{}^*$ the concentration in the convective zone,

Table 2
Depth of inversion of the meridional circulation, surface circular frequency and equatorial velocity.

M^*	$\rho(\mathrm{g\ cm^{-3}})$	$\Omega(\mathrm{s^{-1}})$	$v_{equ}(\mathrm{km\ s^{-1}})$
1.2	6.18E-3	5.09E-5	38
1.1	5.55E-2	1.52E-4	102
1.0	1.85E-1	2.78E-4	167
0.9	4.46E-1	4.32E-4	232

we have, using the elementary model of Schatzman (1981) and Schatzman and Maeder (1981),

$$X_7{}^* = X_7 \frac{K_F\ \Omega^{4/3}}{4\pi\ R^4 \left(\frac{r}{R}\right)^2 \rho\ H_P\ p + K_F\ \Omega^{4/3}} , \tag{5.18}$$

where $-K_F\ \Omega^{7/3}$ is the rate of loss of angular momentum. For $M^* = 0.9$ with $T_{\rm eff} = 5140$ K, $H_P = 4.21\ 10^9$ cm, $T_1 = 2.33\ 10^6$ K, $\rho = 0.41$ g cm^{-3}, one finds $p = 4.63\ 10^{-14}$ s^{-1} for an equatorial velocity of 200 km s^{-1} or $\Omega = 3.71\ 10^{-4}$ s^{-1}, and $(X_7{}^*/X_7) = 0.57$; which is quite a reasonable value.

From these quantities, it is possible to calculate the mass which has to be transferred from the disc to the star in order to satisfy the condition of conservation of angular momentum. It is about 0.003 solar masses and is quite compatible with the order of magnitude given by Bertout *et al* (1988).

After the disappearence of the disc, the spin down process can start. There must be a very rapid spin down of the layers immediately below the convective zone, in such a way that the turbulent diffusion coefficient drops quickly, otherwise there would be no lithium left in these fast rotators. The propagation of the spin-down from the outside to the inside is certainly a characteristic of initially fast rotating stars. In the case of the Sun, this suggests, deep inside the Sun, a possible memory of a high initial angular velocity (Vigneron *et al*, 1990). When looking at the neutrino problem, this should be kept in mind.

This introduces a new parameter, the initial angular velocity when arriving on the main sequence. If we look at the expression of $\omega(t)$ given in section 3, it is clear that the asymptotic value of the angular velocity for $t >> t_0$ is independent of ω_0. This can explain the very small dispersion of the periods of rotation of the late spectral types of the Hyades, for which the condition on t is fullfilled.

This raises immediately the question of the stars strongly deficient in lithium in α Per. Stars reaching the main sequence with a high angular velocity will destroy quickly their lithium. If we assume, following the preceding results, that a star, during its evolution along the Hayashi track is surrounded by an accreting disc, the final rotational velocity and the lithium abundance will depend on the epoch of disappearance of the disk: before of after reaching the main sequence. The slow rotators and lithium deficient stars of α Per could very well be stars which have lost their disk a short time before reaching the main sequence and had time to spin down, at least in their outer layers, and to burn their lithium.

6. Conclusion.

We have used extensively here the turbulent diffusion coefficient given by Zahn (1983 a, b). This expression is obtained by assuming a quasi-solid body rotation. It seems to me that the time dependent turbulent diffusion mixing is well established, and that the dependence in $t^{-3/4}$ is acceptable. This provides a consistent picture of the time dependence of the period of rotation, of the mass dependence of the period of rotation for a set of stars of the same age (Rebolo and Beckman, 1988) and of the lithium abundance. If we look more carefully to the whole set of data (as collected by A. Baglin and Y. Lebreton, 1990) we find that if the general tendency is explained, there are lots of details which do not enter in the picture. We can make a short list of the standing problems : (1) the depth of the convective zone depends on the chemical composition. What is the dependence of the lithium abundance on the abundance of metals ? (2) What is the exact definition of the lower boundary of the fully mixed region ? The transition from the high value of the turbulent diffusion coefficient in the convective zone to the low value in the differentially rotating radiative zone should take place over a very short distance. This is possible with high Rayleigh numbers (Zahn *et al*, 1982) but the exact depth of penetration, through the value of $\Delta\nabla$, depends on the chemical composition and on the temperature. (3) When the accretion disk stops providing angular momentum to a fast rotator, the surface convective zone is the subject of a strong electromagnetic braking. How does the loss of angular momentum propagate inside the star? The shear flow due to a high angular velocity gradient generates a turbulence which determines the flux of angular momentum from the fast rotating inside to the already slow rotating outside. What kind of closure conditions should we write in order to express this feed back mechanism ? *With this last question we are back to the very subject of this workshop.*

References

Baglin, A., 1988, private communication.

Baglin, A., Morel, P. and Schatzman, E. 1985, *Astron. Astrophys.* **149**, 309.

Baglin, A. and Lebreton, Y. 1990, in Inside the Sun, G. Berthomieu and M. Cribier Eds., Kluwer, p. 437.

Balachandran, S., Lambert, D. L. and Stauffer, J. R. 1988, *Astrophys. J.* **333**, 267.

Benz, W., Mayor, M., and Mermillod, J. C. 1984, *Astron. Astrophys.* **138**, 93.

Bernas, R., Gradsztajn, E., Reeves, H. and Schatzman, E. 1984, *Ann. Phys.* **44**, 426.

Bertout, C., Basri, G. and Bouvier, J. 1988, *Astrophys. J.* **330**, 350.

Bodenheimer, P. 1965, *Astrophys. J.* **142**, 451.

Boesgaard, A. M. 1987, *Astrophys. J.* **321**, 967.

Boesgaard, A. M. and Tripico, M. J. 1986, *Astrophys. J.* **302**, L49.

Boesgaard, A. M., Budge, K. B. and Burck, E. E. 1988a, *Astrophys. J.* **325**, 749.

Boesgaard, A. M., Budge, K. B. and Ramsay, M. A. 1988b, *Astrophys. J.* **327**, 389.

Bohugas, J., Carrasco, L., Torrès, C. A. O. and Quast, G. R. 1986, *Astron. Astrophys.* **157**, 278.

Butler, R. P., Cohen, R. D., Duncan, D. K. and Marcy, G. W. 1987, *Astrophys. J.* **319**, L19.

Canal, R. 1974, *Astrophys. J.* **189**, 531.

Cayrel, R., Cayrel de Stropbel, G., Campbell, B. and Däppen, W. 1984, *Astrophys. J.* **283**, 305.

D'Antona, F. and Mazzitelli, I. 1984, *Astron. Astrophys.* **138**, 431.

Duncan, D. K. and Jones, B. F. 1983, *Astrophys. J.* **271**, 663.

Durney, B. R. and Latour, J. 1978, *Geophys. Ap. Fluid Dynamics* **8**, 241.

Dziembowski, W. A., Goode, P. R. and Libbrecht, K. G. 1989, *Astrophys. J.* **337**, L53.

Foshina, M., Martins, J. B. and Tavares, O. A. P. 1984, *Radiochim. Acta* **35**, 121.

Garcia Lopez, R. J., Rebolo, R. and Beckman, J. E. 1988, *Publ. Astrn. Soc. Pacific* **100**, 1489.

Gradsztajn, E. , 1965, *Ann. Phys.* **10**, 791.

Heading, J. 1982, An introduction to Phase-Integral Methods, Methuen and Co LTD, London.

Hobbs, L. M. and Pilachowski, C. 1986a, *Astrophys. J.* **309**, L17.

Hobbs, L. M. and Pilachowski, C. 1986b, *Astrophys. J.* **311**, L37.

Kraft, R. P. 1987, *Astrophys. J.* **150**, 551.

Kraft, R. P. 1969, in Stellar Astronomy, H. U. Chin, R. T. Warasila, F. L. Rem. Eds, Gordon and Breach Science Publishers N.Y., N.Y., p. 315.

Kocharov, G. E. 1988, *Astrophysics and Space Science Reviews* **8**, 313 (Section E of Soviet Scientific Reviews).

Lamb, H. 1932, Hydrodynamics, Camb. Univ. Press.

Léorat, J. 1975, Thèse d'Etat, Université de Paris VII.

Mazzei, P. and Pizatto, L. 1989, *Astron. Astrophys.* **213**, L1-L4.

Mestel, L. and Spruit, H. C. 1987, *Mon. Not. Roy. Astron. Soc.* **226**, p 57.

Moffatt, H. K. 1978, Magnetic Fluid Generation in Electrically Conducting Fluids, Camb. Univ. Press.

Moffatt, H. K. 1984, in Champs Magnétiques Stellaires, A. Baglin Ed., Société Française des Spécialistes d'Astronomie, Paris 1984, p. 253.

Parker, E. N. 1975, *Astrophys. J.* **198**, 205

Parker, E. N. 1977, *Astrophys. J.* **215**, 370.

Pinsonneault, M. H., Kawaler, S. D., Sofia, S. and Demarque, P. 1989, *Astrophys. J.* **338**, 424.

Proffitt, C. R. and Michaud, G. 1989, *Astrophys. J.* **346**, 976.

Rebolo, R. and Beckman, J. E. 1988, *Astron. Astrophys.* **201**, 267.

Roxburgh, I. W. 1983, in Solar and Stellar Magnetic Fields, J.O. Stenflo Ed., Reidel, Dordrecht, p. 449.

Schmitt, D. and Schüssler, M. 1989, *Astron. Astrophys.* **223**, 343.

Sackman, I. J., Smith, R. L. and Despain, K. H. 1974, *Astrophys. J.* **187**, 555.

Schatzman, E. 1962, *Ann. d'Ap.* **25**, 18.

Schatzman, E. 1981, in Turbulent diffusion and the solar neutrino problem, CERN 81-11.

Schatzman, E. 1984, in Observational Tests of Stellar Evolution Theory, IAU symposium N 105, A. Maeder and A. Renzini Eds., Reidel, Dordrecht, p. 491.

Schatzman, E. 1989, in Turbulence and Nonlinear Dynamics in MHD Flow, Proceedings of the workshop on Nonlinear Dynamics in MHD flow, Cargèse, France, July 4-8, 1988, M. Meneguzzi, A. Pouquet, P. L. Sulem Eds., North Holland, Amsterdam, p. 1.

Schatzman, E. 1990, in Inside the Sun, G. Berthomieu and M. Cribier Eds., Kluwer, p. 5.

Schatzman, E. and Maeder, A. 1981, *Astron. Astrophys.* **96**, 1.

Schatzman, E. and Ribes, E. 1987 in New and Exotic Phenomena, O. Fackler and J. Trân Than Vân Eds., Editions Frontières, Gif sur Yvette, p. 365.

Schüssler, M. 1977, *Astron. Astrophys.* **56**, 439.

Schüssler, M. 1979, *Astron. Astrophys.* **71**, 79.

Schüssler, M. 1983, in Solar and Stellar Magnetic Fields, Origins and Coronal Effects, J. O. Henfls Ed., D. Reidel, Dordrecht, p 213.

Shu, F. H., Lizano, S., Ruden, S. P. and Najita, J. 1988, *Astrophys. J.* **328**, L19.

Simon, Th., Herbig, G. and Boesgaard, A. M. 1984, *Astrophys. J.* **293**, 551.

Skumanich, A. 1972, *Astrophys. J.* **171**, 565.

Soderblom, D. R. 1983, *Ap. J. Suppl. series* **53**, 1.

Soderblom, D. R. and Stauffer, J. R. 1984, *Astronom J.* **89**, 1543.

Spite, F., Spite, M., Peterson, R. C. and Chaffee, F. H. Jr. 1987, *Astron. Astrophys.* **171**, L8.

Stauffer, J. R. *et al*, 1988, preprint.

Vigneron, C., Catala, C., Mangeney, A. and Schatzman, E. 1990, in Inside the Sun, G. Berthomieu and M. Cribier Eds., Kluwer, p. 513. Posters to be published in a special issue of Solar Physics.

Zahn, J.P. 1983a, in Astrophysical Processes in Upper Main Sequence stars, 13th Advanced Course, Swiss Society of Astronomy and Astrophysics, SAAS-FEE, 1983, B. Hauck, A. Maeder Eds., Geneva Observatory, CH-1290, Sauverny, Switzerland, p. 253.

Zahn, J.P. 1983b, in Instabilités Hydrodynamiques et Applications Astrophysiques, A. Baglin Ed., Société Francaise des Spécialistes d'Astronomie, Paris 1983, p. 389.

Zahn, J.P. 1988, Ecole d'été des Houches (à paraitre).

Zahn, J.P., Toomre, J. and Latour, J. 1982, *Geophys. Ap. Fluid Dyn.* **22**, 159.

Zahn, J.P., 1989, private communication.

THE OBSERVATIONAL FACTS

SURFACE ABUNDANCES

G. Cayrel de Strobel
Observatoire de Paris, Section de Meudon
F-92195 Meudon Cedex, France

Abstract. A general view of stellar surface abundances is presented. Chemical composition differences are discussed in the light of differences in internal structure, in stellar atmospheres and in nucleosynthesis from star to star. An effort has been made to discuss non solar elemental abundances in stars as :

i) caused by physical processes in the interior or on the surface of stars

or, as :

ii) coming from differently metal enriched interstellar material.

Some examples of interesting abundance results are given.

1.Introduction

The improvement of the results of elemental abundance determinations in stellar atmospheres with the advent of very performant telescopes, spectrographes, solid state detectors and reliable model atmosphere computations, has shown that stars having exactly the same chemical composition of the Sun are rare, if they exist at all. However, the habit has been taken to call "normal" the surface elemental abundances of the Sun and to refer to them in determining the abundances of other stars.

Stellar surface abundances deviating from the solar ones, can be explained in two ways : either they differ from the Sun because the stars were formed from an interstellar matter, which chemical composition was different from that of the Sun, or because stellar surface abundances have been altered by physical processes as : selective mass-loss, radiative diffusion modified or not by magnetic fields, nuclear burning induced by convection and turbulent mixing. This latter type of surface abundance peculiarities provides important clues on the internal structure of a star. For example the abundance stratifications on the surface of magnetic Ap-stars tell us about the role played by their magnetic field and the depletion of lithium along the main sequence of open clusters tell us about the thickness of the convective zone in each of the analysed main-sequence stars.

In what follows, we shall mainly discuss chemical composition differences from star to star, induced by physical processes in their atmospheres. Then we shall proceed to go over to the chemical composition of low-mass, disk and halo stars. The abundances of these objects can be interpreted to a great extent with the help of nuclear processes which took place inside massive stars and whose products were returned to the interstellar matter, (from which the small mass stars were then formed), mostly during the explosion of supernovae. Small mass stars have a developped convective zone which prevents the formation of abundance particularities at their surfaces.

Spectroscopic abundances researches rely more and more on high resolution, high S/N spectra taken with solid state detectors. The increase in accuracy has sometimes been by a full order of magnitude (R. Cayrel 1988).

All along this talk we have to keep in mind that stellar abundance determinations, which we can trust, are only feasible if we have previously determined with great care the fundamental physical parameters of the stars : effective temperature, surface gravity, velocities fields etc. Errors in the estimated fundamental parameters introduce abundance errors ≥ 0.1 dex (Gustafsson 1988). The errors are greatest for stars of spectral type O, M, N, and greater for supergiants than for giants and dwarfs. Errors in model atmosphere lead to errors in abundance determinations of about the some order of magnitude as those caused by other sources. It is therefore, for the knowledge of the chemical composition of a star, as important to improve the determinations of T_{eff} and log g as the model atmosphere computation.

We will now present abundance results picked up, very subjectively, from a wealth of interesting researches. We regret only not being able to comment on many more. We shall begin to discuss abundances in high mass stars, and then proceed towards less massive stars.

2. Chemical abundances in massive stars

2.1 *O and Of stars.*

Recently, several papers have appeared on these stars, deepening the knowledge of their physical properties. Abbott and Hummer (1985), Bohannan et al (1986), Bohannan et al (1988) Voels et al (1989) have obtained high S/N H and He line profiles of O-stars hotter than 35000K. These profiles have been analysed with a technique of high precision line profiles fitting, using wind-blanketed model atmospheres. The effect of back-scattering of radiation into the photosphere from the stellar wind was first considered by Hummer (1982) who called this mechanism "wind-blanketing". Wind-blanketing reduces the effective temperatures needed to produce photospheric absorption lines of a given strength. This reduction can be as large as 10000K for very luminous O-type stars. Effective temperature, spectroscopic gravity, and He-abundances have been determined for some early type stars by the above authors and are given in Table 1, together with other stellar parameters. The estimated error in T_{eff}, is ± 1500K at spectral type O4, and ± 1000K at O9.5; in gravity ± 0.1 dex at O4 , and ± 0.5 dex at O9.5; in He-abundance by atom number per H-atom: N(He)/N(H), ± 0.03. The large He abundance observed for ζ Pup and α Cam is consistent with the high degree of evolution of these stars, which have lost some 40 $\mathbf{M}_\odot$ of material (≈50%) from their original mass of 70 - 90 $\mathbf{M}_\odot$. Very up-to-date models of stellar evolution predict that these very evolved stars are already turning back toward the hotter part of the HR-diagram (Maeder 1983).

Table 1: O, Of Stars

Name HD	V B-V	Sp	Vsini km s^{-1}	$\log L/L_\odot$	Teff K	log *g*	N(He)/N(H)
ζ Pup 66811	2.25 -0.26	O5 Iaf	211	6.0	42000	3.5	0.17
α Cam 30614	4.29 +0.03	O9.5 Iae	95	5.8	30000	2.9	0.20
ξ Ori A 37742	2.05	O9.5 Ibe	140	5.8	31000	3.2	0.10
δ Ori 36486	2.23 -0.22	O9.5 II	152	5.6	33000	3.4	0.10
9 Sgr 164794	5.97 0.00	O4 V	140	6.1	46000	3.9	0.10

2.2 *Wolf-Rayet (WR) stars*

These stars are massive and exhibit in their spectra strong emission lines. Their abundance anomalies are consequently determined from emission lines. For this highly evolved stars the modelling of CIV and NIV lines causes important constraints on evolutionary models. In the CNO cycle the C and the O are converted to N and C on a very short time-scale (Maeder 1983). In the WR stars the results of nuclear fusion in the core are observable on their surface because the upper layers of these stars have been stripped away and their inside become visible. The WR star, HD 508996, is a good example of a star showing the products of central nuclear combustion on its surface. A detailed analysis of the spectrum of HD 50896 (Hillier 1988) has found that it is overabundant in N and underabundant in C. The C/N abundance ratio of HD 50896 is equal to 0.07 (within a factor of 3), as against the Sun in which C/N = 1.06 (Anders and Grevesse 1989).

2.3 *He-strong and He-weak stars*

These stars are hot main sequence stars with anomalously strong or weak He-lines for their effective temperature. Many members of the two classes have strong magnetic fields and represent high temperature extensions of Ap-stars. Bohlender et al (1987) have determined abundance and surface magnetic field geometry for several He-peculiar stars with large, well determined magnetic fields. The prototype of a He-strong star is σ Ori E. With an effective temperature of 25000K, σOri E has an He-abundance by number of : N(He)/N(H)=0.40 ± 0.05; it has also strong magnetic field variations from + 2800G to -1500G. In the very young *h* and χPer clusters, B stars are He-deficient. (Wolff and Haesly 1985).

3. **Interesting abundance results in A stars.**

3.1 *Do normal A stars exist?*

In focusing our attention towards less massive stars, the A stars, we are struck by the variety of chemical compositions existing among them. It is very difficult, if not impossible to compile a list of A stars displaying solar elemental abundances. The majority of such stars, when analysed in detail with the help of high S/N, high resolution spectroscopic material, shows very conspicuous abundance anomalies. Two of the brightest early A type stars in the sky, Sirius and Vega with almost the same T_{eff}, and not very different gravity, luminosity, rotational velocity, and Vsin i have been found, the first, - metal-rich by + 0.75 dex, and the second, metal poor by –0.60 dex. Table 2 shows some physical parameters of these stars together with oPeg one of the few metal-normal A-stars. If, we reject the idea that the difference in Fe-content between Sirius and Vega comes from different metal enrichment of the parent interstellar matter, then we have to interpret this difference as a combination of several physical processes. Most of the anomalous chemistry in A stars is explicable in terms of diffusive fractionation, due either to gravitational settling or to radiative support. General problems related with stellar magnetism are not restricted to Ap stars alone, and hydro-magnetic processes acting during star formation, pre-main-sequence, as well as during the H-burning-phase have to be carefully considered in the interpretation of elemental abundances of A stars.

Table 2 Two famous early A stars and one [Fe/H] solar (normal) A star.

Name HD	V B-V	Sp	Vsini kms^{-1}	$\log L/L_\odot$	T_{eff} (K)	logg	$[Fe/H]_\odot$
Sirius 48915	-1.46 0.00	A1 V	+13	1.34	9700	4.3	+0.75±0.15
Vega 172167	0.03 0.00	A0 V	+14	1.70	9600	3.9	−0.60±0.15
o Peg 214994	4.79 -0.01	A1 IV	+12	1.28	9500	3.5	0.00

3.2 *Non-LTE analysis and hydrodynamic stability of the atmosphere of Vega*
A very interesting detailed analysis of Vega (HD 172167) has been carried out by Gigas (1986,1988a,1988b) with the help of NLTE-line-computations. Gigas has recalculated the abundances of three elements : Mg, representing the "light elements", Fe, the "iron peak" elements and Ba, the "heavy elements". The results indicate marked underabundances of these elements relative to the solar values.

Magnesium	$[Mg/H]_{\odot}$	=	(-0.58 ± 0.15) dex
Iron	$[Fe/H]_{\odot}$	=	(-0.55 ± 0.22) dex
Barium	$[Ba/H]_{\odot}$	=	(-0.21 ± 0.23) dex

The NLTE $[Fe/H]_{\odot}$ value of Vega is in good agreement with that obtained through classical LTE models. (Cayrel de Strobel et al. 1985).
Gigas has also confirmed that Vega has a higher microturbulence than predicted by the mixing-length theory, in assuming that the atmospheres of A0 type stars are static. He found that the microturbulence of Vega is ≈ 2.0 Kms^{-1} against 0.25 - 0.50 Kms^{-1} as predicted by mixing-length. To find a possible explanation for this difference in microturbulence, Gigas (1988) is working on numerical simulations of convective phenomena in A-type-stellar-atmospheres. He employes the method of "bicharacteristics", which solves the time dependent non linear equations of motions in two dimensions on the assumption of cylindrical symmetry. First preliminary results indicate the presence of atmospheric oscillations, with flow velocities considerably larger than those predicted by the mixing-length theory and comparable to the observed microturbulence.

The very typical and seemingly well known A0 type star, Vega, employed in these last 50 years as a primary standard for flux measurements, as a reference star for abundance studies, as a photometric and spectral type standard in a great quantity of researches, is a good example of how poorly the physical status of its atmosphere and interior was known until now. High S/N, high resolution material, NLTE computations, hydrodynamics more elaborated than the mixing-length theory have been necessary to advance the knowledge of the physical structure of Vega.

3.3 *Ap and Am stars*
Abundance anomalies in high and intermediate-mass stars can be found in a large portion of the HR diagram. They spread from the upper main-sequence B-stars to the midle main-sequence F-stars, up to the horizontal-branch and down to the white dwarfs. Much of the anomalies of Ap and Am stars are explainable by radiative diffusion processes, but the detailed chemical composition of these stars cannot be understood without consideration of several other hydrodynamic and magnetohydrodynamic processes. For example, high mass-loss rates are capable to explain the early metal deficient, λ Boo stars, while somewhat lower rates may produce the anomalies which appear in Am-stars. Additional physical factors that may play an important role in surface chemical peculiarities are : stellar rotation, dynamo, electromagnetic braking, meridional circulation, turbulent mixing, (Alecian and Vauclair 1983, Schatzman 1989, Zahn 1989).

The magnetic Ap-stars show magnetic field variations accompanied by spectral variations caused by the inhomogeneous distribution of elements on the stellar surface. It is believed that the magnetic field plays an important role in determining this distribution. Concerning the radiative diffusion, it has been predicted theoretically (Michaud et al. 1981), that during the main-sequence phase of the star, the magnetic-field lines guide the chemical elements in the stellar atmosphere in places where they accumulate.
Although many Ap stars have strong magnetic fields, the so called Hg-Mn stars have non detectable fields. These stars are interesting for the study of the abundance patterns that develop in non-magnetic Ap-stars.

The elemental abundance in Bp and Ap-stars vary with T_{eff} : Mn-stars have higher temperatures than Eu-Cr-Sr-stars. Some Mn-stars have highly enhanced Hg (by factors of more than 10^6 with respect to the Sun). A study of the gallium diffusion has been made by Alecian and Artru (1987, 1988) using up-to-date GaII atomic data. The resonance line at 1414.40 Å in

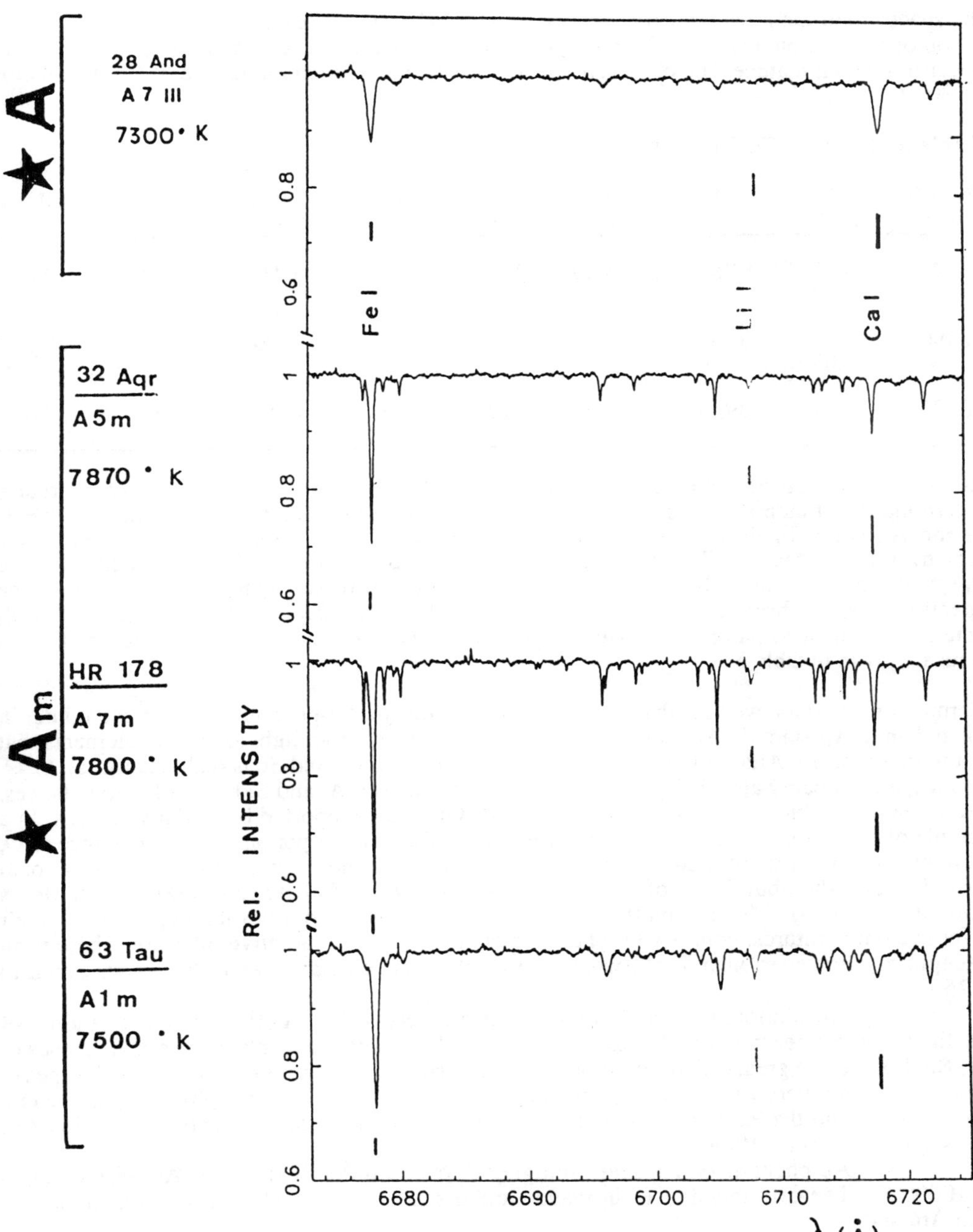

Fig. 1. CFHT high S/N Reticon spectra (Burkhart and Coupry, 1989) of one A and three Am stars.Note the intensities of the FeI line (6678.00 Å), and the CaI line (6717.69 Å) in the A star and in the Am stars. The two lines have almost the same excitation potential.

the spectrum of the most typical Hg-Mn star : HD 175640, has been computed in LTE with an inhomogeneous abundance of Ga throughout the atmosphere. The computation of the synthetic spectrum which matches the observed one was done with an overabundance of Ga by a factor of 3000.

Table 3: Si-Sr-Cr-Eu-Hg Ap stars , Bp stars

NAME HD	V B-V	Sp	Vsini kms^{-1}	T_{eff} K	logg	$\lvert Be \rvert$ Gauss	$[Fe/H]_\odot$	$[Si/H]_\odot$
13329	6.37 -0.14	B9pSiSrCr	<20	10100	4.5	3300	+0.60	+1.80
224801	6.38 -0.07	B9pSiEu	49	12000	3.7	2300	+0.54	+1.60
18296	5.11 -0.01	B9pSi	19	11200	3.0	1350	0.00	+1.20

Reliable elemental abundance determinations in stars require, first of all, a very precise determination of their effective temperature. This is especially true for such complicated objects as the Ap-stars. In these stars the overabundances of Si and other elements result in a redistribution in their visible spectrum of the flux absorbed in their UV. Consequently, the T_{eff}'s of Ap-stars cannot be determined from their optical fluxes alone. In this respect some specific T_{eff}-calibrations have been elaborated by Megessier (1988). Some very well determined physical parameters (Megessier 1989 private communication) of three Si-B_p-A_p stars are given in table 3.

Coming to Am-stars, we note that the elemental overabundance-phenomenon is much milder in them than in Ap-stars. In Am-stars overabundances increase from light to heavier elements, but not monotonously. Among light elements the deficiency of Ca and Sc is real (see fig. 1); C, O, and Mg are probably also deficient, but at the same time Na, Al and Si have solar abundances. From theory we know that the infrared triplet of OI is not supported by radiative forces. In a sample of Am stars, C. Van't Veer et al. (1990) have found an important underabundance of O, in agreement with theory. Burtkhart et al (1988) have found on high quality observational material, that the abundance of Li in Am stars having T_{eff}'s between 7000 and 9000K is normal : log N(Li)=3.0 dex, (on the scale log N(H)=12 dex) and does not depend upon their effective temperatures, contrary to what predicts the theory of radiative diffusion.There is an exception , however, near 7850, where the Li-abundance of the star 16 Ori goes down to 0.75 dex.

High resolution, high S/N observations are needed for a definitive understanding of the light-element behaviour in Am star. In regard to the heavier elements the overabundances of Sr, Ba, Eu, Gd are greater than those of Zr, Ce and Nd. The overabundances of the "Fe-peak" elements are not very important. In table 4 are given, together with other physical parameters, the $[Fe/H]_\odot$ and the $[Ca/H]_\odot$ abundances of four of the most classical and best studied Am stars. (Burkhart *et al.* 1989).

The Ap phenomenon appears at higher T_{eff}'s (>10000 K) than the Am phenomenon (<10000 K). The overabundances in magnetic Ap-stars are larger than in non-magnetic Ap and Am-stars.

Specialists of Ap and Am stars know that it is very difficult to classify such stars per given abundance-groups. Each chemical peculiar A star has its own abundance finger-prints. If we want to interpret the individual elemental abundance in Ap, Am-stars, we have to consider more than one physical process and see, first, their contribution to the deviations with respect to the normal solar-abundances. The competitions (G.Vauclair and S. Vauclair 1978) between these various processes does, then, determine the quantitative patterns of abundances, but the detailed theoretical prediction of those is still very rarely done, because explaining one by one the different abundance anomalies in each star is not a simple undertaking and requires the most rigourous input-physics known nowadays.

Table 4 Am stars

Name HD	V B-V	Sp	Vsini ξ_t	T_{eff} K	log$g$	[Fe/H]$_\odot$	[Ca/H]$_\odot$
63 Tau 27749	5.6 0.30	A1m	10 3.8	7520	4.5	+0.5	-0.6
68 Tau 27962	4.3 0.05	A2IV	18	9000	4.0	+0.5	+0.1
81 Tau 28546	5.5 0.26	Am	21 4.0	7750	4.0	+0.5	-0.1
16 Ori 33254	5.4 0.24	A2m	15 3.0	7880	4.0	+0.7	-0.3

3.4 *Two of the most exotic specimen of chemical peculiar stars*

3.4.1 The rare earth star : HD 101065
This star is also called "PRZYBYLSKI's star" after the astronomer who has discovered it. (Przybylski 1977) . The visible part of its spectrum is mostly constituded by lines of rare-earths (mostly Holmium): some lines of the iron-peak elements are visible but exceedingly faint. Recent IUE spectra of HD 101065 show many, and sometimes, strong lines of FeII, CrII and TiII (Wegner 1983).
From the energy distribution of HD 101065 determined on low resolution UV spectra, Wegner attributed to the star a Teff of about (10000 to 12000 K) much higher than that obtained from its visible spectrum (Teff=6000 K). This indicates that the severe blanketing of the star prevents in the visible a reliable energy distribution determination for an accurate Teff calibration. In addition to its very abnormal spectrum, (specialists say that it is the most abnormal A-type spectrum known), HD 101065 has a magnetic field of: 2100–2500 G (Wolff and Hagen 1976) and 12 minutes light variations, in common with many other rapidly oscillating Ap stars.

3.4.2 *The variable Ap star : FG Sagittae*
This star is probably in the process of displaying on its surface the products of internal nucleosynthesis, more specially those built up with the s-process (Cowley et al 1985). The weakness of the Fe-lines and of other lines of the "iron-peak" makes it unlikely that this object will become a BaII-star. Following Cowley et al (1985) the "iron-peak" lines of FG Sge have weaked markedly, while lines of BaII and other lanthanides have strengthened. There is an exciting possibility that if the lanthanides continue to strengthen, and the "iron-peak" elements to weaken on the surface of FG Sge, its spectrum could become very similar to that of Przybylski's star!

4. The exciting role of F, G and K stars in Astrophysics

4.1 *F, G and K stars as stellar population tracers.*
The role played by F, G and K stars in Astrophysics is well known. These stars are not only used for intrinsic detailed analyses of their atmospheres and of the modeling of their interiors, but they are used to study stellar populations, the abundance gradients across the galactic disk, the constraints on nucleosynthesis imposed by the chemical composition of extremely metal deficient objects, the connection between kinematical, dynamical and chemical evolution of our Galaxy and of other galaxies. This is because it is only from late F or early G type that evolution has not depleted the initial stellar population of the galaxies, and that the full span of

stellar ages is still present. Solar type stars, and later, are composed by a great variety of objects. Some of them are massive young supergiants, crossing rapidly the F, G and K-star region of the HR diagram, but the majority of them are small mass stars, evolving slowly. Their Teff's go from about 6300 K to about 4000 K. Their ages (Carney *et al.* 1989) span from now to the beginning of the Universe, whenever it began (bets go from 8 to 18 Gyrs).

4.2 *Metal enrichment in the Galaxy from the chemical composition of F, G and K stars.*

F,G and K stars belong to three different galactic subsystems (Gilmore and Reid 1983, Gilmore and Wyse 1985). Following these authors, taking as metal abundance indicator the well known parameter : $[Fe/H]_{\odot}$ we can define F, G, K, stars as:

Halo stars	with		$[Fe/H]_{\odot} < -1.0$
Thick Disk stars	"	$-1.0 <$	$[Fe/H]_{\odot} < -0.4$ dex
Thin Disk stars	"	$-0.4 <$	$[Fe/H]_{\odot}$

However, it must be noted that if these three groups are defined kinematically, and not chemically,each one has a spread in metallicity and there is some overlapping in their metallicity distribution (Laird et al 1989).

Several reviews have been dedicated recently to F, G and K stars (G. Cayrel de Strobel and Bentolila 1983, Spite and Spite 1985, Grenon 1987, Gustafsson 1988, Lambert 1988, Wheeler et al. 1989).

A wide range of metallicities is observed among Halo-stars: $-4.5 < [Fe/H] \lesssim -1.0$ dex (Bessell and Norris 1987). The non-evolved Halo-stars have extended convective zones preventing the formation of peculiar surface abundances. Except, for lithium the abundances of the α, the "iron-peak", and the s and r elements found in these stars, give then informations of the nuclear processes which have occured before their birth. SN-ejectae, as well as, mass flows from stellar winds have thus fed up with newly synthetized elements the parent interstellar medium of low mass, long living, Halo-stars.

New, more precise analyses of Halo-dwarfs have shown relative abundance variations of particular elements $[N_{el}/Fe]_{\odot}$ respect to $[Fe/H]_{\odot}$. Good reviews, illustrated by convincing diagrams describing improvements in relative elemental abundances, have been written by Lambert (1987, 1989) and by M. Spite (1990) during the last few years.

In very metal deficient Halo-stars, O is overabundant with respect to Fe, whereas the s-process elements Sr, Y, Zr and Ba are underabundant in respect to Fe. The overabundances of O with respect to Fe is explained (Matteucci and Tornambe 1985) by the time delay between production of O and Fe by supernovae of type II and I; The population II is apparently fed exclusively by supernovae of type II, whereas pop I has accumulated enrichment by the two types of supernovae. Because supernovae of type I produce a lot of iron, the O/Fe ratio goes down where there is a significant contribution of type I SN to the nucleosynthetic yields. The underabundance of s-elements (Sr, Y, Zr, Ba) in halo dwarfs has a different explanation. Because s-elements are supposed to be produced by asymptotic giant branch (AGB) stars, (during "dredge-up" of burned matter mixed with unburned layers), having already iron-peak elements, they need, to be produced, a first generation of stars producing iron-peak elements, plus a second one producing the "dredge-up". Therefore the production of s-elements (secondary elements) is delayed with respect to O and Fe production.

The enrichment in heavy elements in the thick Galactic Disk has not been uniform neither in time, nor in location, in our Galaxy. Here too, as in the case of Halo-stars, the enrichment was tightly bound to the nature of the nucleosynthesis which has acted in supernovae or at the center of massive stars enriching the parent clouds at the epoch in which the disc-stars were formed. We know that our nearest neighbours, the field stars, are chiefly old disk population stars, sometimes quite older than the Sun. Their metallicities vary from factors by 1 to 3, more, to factors by 1 to 5 less, than that of the Sun. Recent progress in spectroscopy of solar type stars has been reviewed by Nissen (1990). In this review particular attention has been paid to the determination of stellar atmospheric parameters and the abundances of selected elements. A search for real solar twins has been undertaken by Cayrel de Strobel and Bentolila

(1989).For the moment these authors have not yet found an exact solar twin which matches the Sun's mass, age, chemical composition, effective temperature, gravity, chromospheric activity etc., etc. Metal abundance gradients of disc-stars as a function of their galactocentric distance and as a function of time have been proposed by different authors, (Twarog 1980, Grenon 1987).

The metal content of the thin Galactic Disk has been studied with the help of F, G, and K– dwarfs belonging to young galactic clusters. Recently Cayrel et al. (1985, 1988) have analyzed in detailed G and K dwarfs of the Pleiades, Ursa Major, Coma Berenices and Hyades. They have found that the mean metallicities of the stars in the first three clusters do not deviate significatively from solar. The Hyades with $[Fe/H]_{\odot} = +0.12 \pm 0.03$, show a significative enhancement in metals.

Analysing F dwarfs in six galactic clusters, Boesgaard (1989) found that the two oldest clusters Hyades and Praesepe respectively 6.7×10^8 and 7.6×10^8 years old, were metal-enhanced ($[Fe/H]_{\odot} \approx 0.12$), the two youngest clusters, α Persei and Pleiades respectively 2.7×10^7 and 1.5×10^8 years old, were metal-normal $[Fe/H]_{\odot} \approx 0$ and the intermediate age clusters, Ursa Major and Coma, respectively 3×10^8 and 4.3×10^8 years old, slightly metal-deficient. Please, note, that, very recently, the age of the Pleiades has been revised from 6 to 7×10^7 to 1.5×10^8 (Mazzei and Pigatto 1989). The results of Boesgaard indicate that there is no univoque age-metallicity relationship, which signifies that the enrichment in the thin galactic disc has not been uniform on mixing-time scales of less than 10^9 years.

4.3 *SMR Stars*

Are the so-called super-metal-rich (SMR) G and K stars moderately metal enriched (by 0.15 - 0.50 dex), in respect to the Sun, or do they constitue an apart small galactic subsystem, which metal content ranks quite above the Sun ? This is the question which astronomers interested in galactic and extragalactic chemical evolution of stars propound since the classical paper of Spinrad and Taylor (1969).

With the help of the [Fe/H] Catalogue (Cayrel de Strobel et al. 1985), Cayrel de Strobel (1985) has displayed two groups of mild metal-rich field stars. The first one is composed by G and K dwarfs and subgiants, the second one by K-giants. If there is no doubt that the metal enrichment of the first group of stars reflects the chemical composition of the interstellar matter in which these stars were formed, the metal enrichment of the stars of the second group could, at least for some of them (CN-rich), reflect nucleosynthesis produced in their center. Cayrel de Strobel (1987) tried to attribute a "turn-off" age to the slightly evolved stars of her sample of "SMR" stars. She found that such stars have been produced all along the existence of the Galactic Disk, being some very old and some very young. There is an indication that the "SMR" phenomenon was more active in the past than it is now. A possible, but not unique interpretation of this, is that the interstellar medium in our Galaxy became progressively more chemically uniform with time.

Probably the sample of mild "SMR" stars $+0.15 < [Fe/H] < +0.5$ dex contained in the [Fe/H] catalogue, does not constitue a galactic subsystem. At the contrary, the discovery by Withford and Rich (1983) of a group of metal rich stars in the Bulge of our Galaxy, which metallicity is more than 3 times that of the Sun, could make us suppose that in the Bulge of the Galaxy does exist a subsystem of metal rich stars.

5.Lithium an essential element for stellar interior physics

5.1 *Lithium in Halo stars*

Let us finish with lithium this fast review on surface elemental abundances in stars. The abundance results for lithium, either in pop II or in pop I stars, are more and more employed as a precious working tool in the understanding of the multiple physical processes acting in the interiors of the stars. Also, lithium has been, and is used in the study of Big-Bang nucleosynthesis, and has taken a high ranking place among the other Big-Bang nucleosynthesis products (Wagoner 1973, Yang et al 1984, Kawano et al 1988). Spite and Spite (1982a, 1982b) have discovered that Li is present in the hottest halo subgiants or dwarfs ($T_{eff} > 5400$) and that in a large temperature interval, Li/H is constant and independent from their metal

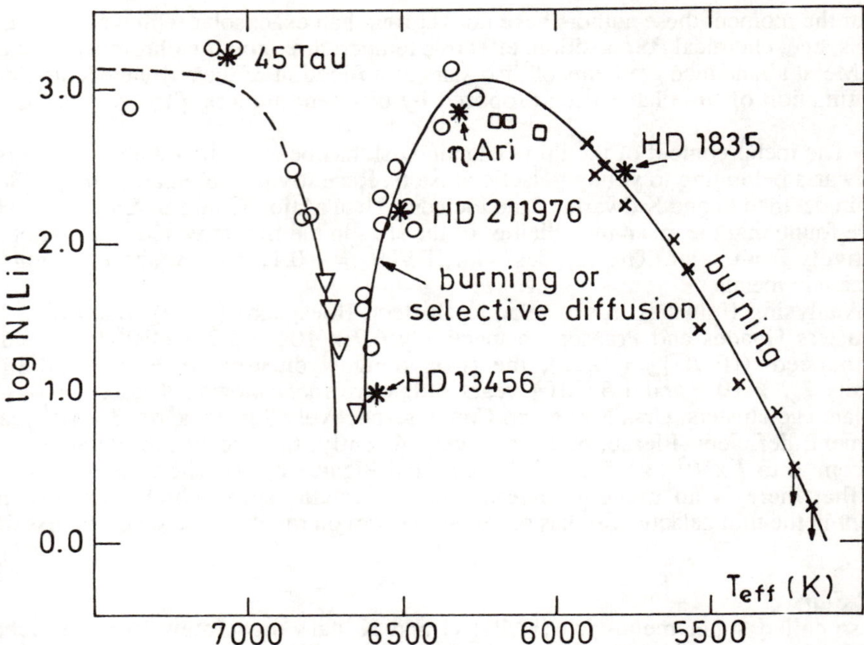

Fig. 2. The Boesgaard - Tripicco chasm of the Hyades and of some field stars : open circles and triangles are from Boesgaard and Tripicco (1986a,1986b),the squares are from Duncan and Jones (1983), the crosses from Cayrel et al.(1984), the asterisks are field stars in which the $^6Li/^7Li$ has been carefully measured by a profil fitting technique (Cayrel et al. 1990).

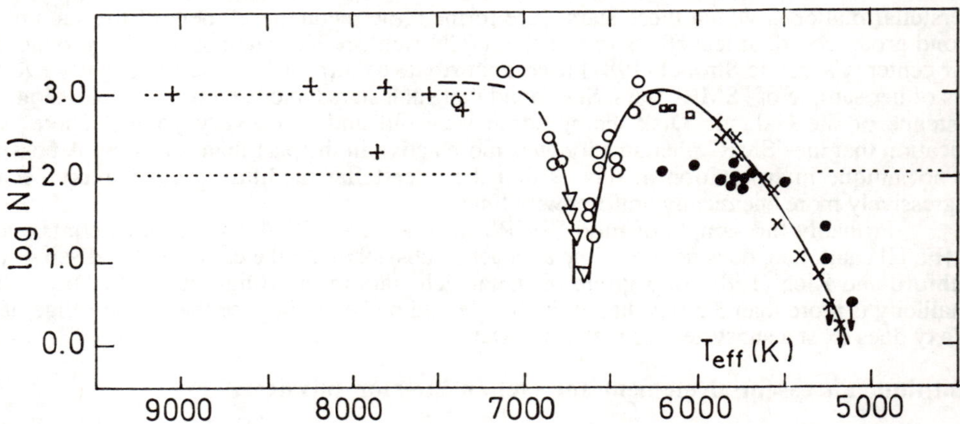

Fig.3. Composite diagram of the Li-temperature relation. In respect to Fig.2, have been added in this diagram Li - abundance values from Am-stars (Burkhart and Coupry 1988) and Li-abundances from hot and cool Halo - dwarfs and subgiants (Spite and Spite 1985). Symbols: same as in Fig.2, in addition: pluses: Am-stars, black circles: Halo - stars. Note, that the constant Li abundancein Halo dwarfs extends on more than 800 K.

deficiency. This is seen in table 5 which reproduces, together with some other physical parameters, the [Fe/H] and log N(^{7}Li) values of three hot halo dwarfs and two cooler ones.

Table 5 Fe and Li abundances in some Halo-dwarfs and subgiants

Name HD BD	V B-V	Sp	T_{eff} K	log*g*	[Fe/H]	logN (^{7}Li)
+20°3603	9.72 0.44	G1VI	6000	4.0	-2.2	2.11
219617	8.17 0.47	G0VI	5660	3.9	-1.4	2.10
94028	8.21 0.48	G5IV	5790	4.0	-1.7	2.09
211998	5.28 0.65	G0V	5200	3.5	-1.5	1.04
Gmb1830 103095	6.49 0.75	G8VI	5140	4.7	-1.4	<0.5

The value log N(^{7}Li) $\approx$ 2.0 dex (on the scale log N (H) = 12.0 dex), found by the Spites in hot halo-dwarfs and confirmed by other authors (Rebolo et al 1986, Hobbs and Duncan 1986) are used by specialists of Big-Bang nucleosynthesis in order to corroborate their computations. But, there is a "caveat" in the observational value of the Li-abundance found in hot halo dwarfs : is this value, the original Li-abundance produced during the Big-Bang explosion, or have these very old halo objects depleted some of their original Li ? If there has been some depletion,the Li-abundance, as found in Halo-dwarfs, cannot be directly compared to the cosmological value (Boesgaard and Steigman 1985).

5.2 *Lithium in disk stars*

The amount of review and scientific papers on lithium appeared in literature, since the pioneering paper of Herbig (1965), is difficult to count. Li is implied in a large deal of problems, the solution of which depends upon high S/N spectroscopy and theories built up with reliable input physics.

In a short, but excellent review on Li, Soderblom (1988) separates what we know, or what we need to know about Li in four sections :

1. What do we really know ?
2. What do we think to know ?
3. What do we not know ?
4. What do we need to know ?

Reading again the paper by Soderblom, 21/2 years after it was presented at IAU Symposium 132 (June 1987), one notices , that, some of the questions or statements in one or another of the four above mentioned sections have been answered or resolved, some of them are now more appropriate for a different section. This is due, thanks to a wealth of new observational results, new theories and fancy interpretations of recent stellar Li-abundances.

Schatzman (1990) and Zahn (1990) in this Conference, Baglin and Lebreton (1990) in the "Inside of the Sun" Conference have presented the highlights of their and other Astrophysicists researches on "the mechanism of Li-depletion", which is the most important question in section 3, in the paper by Soderblom. Balachandran et al. (1988) in the paper

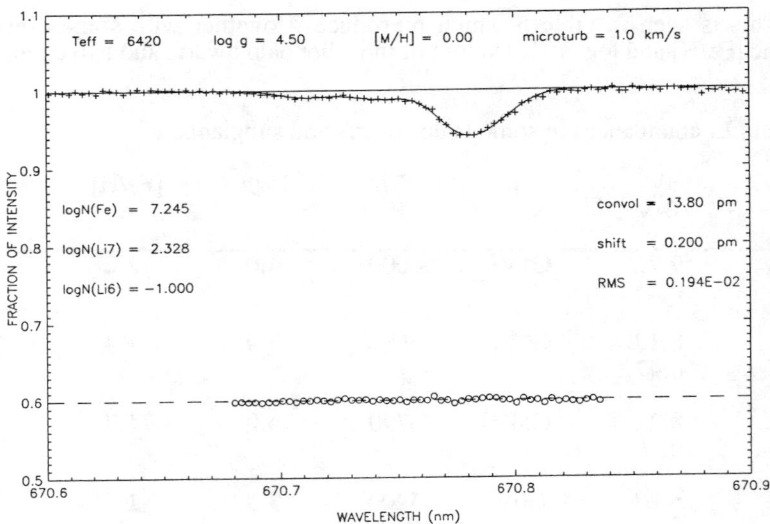

Fig. 4 a. Observed (crosses) and synthetic (full line) spectra of the Li - blend region for HD211976 .The ^{6}Li abundance is set to a negligeable value. The RMS value of the fit is : 0.00194, correlating the photometric accuracy of the observation (S/N≈500). Note that a small shift in wavelength has been applied (0.2 pm or 2 mÅ). This is close to the RMS error on wavelength. The residuals are plotted on the same scale, but with the zero shifted at ordinate 0.6.

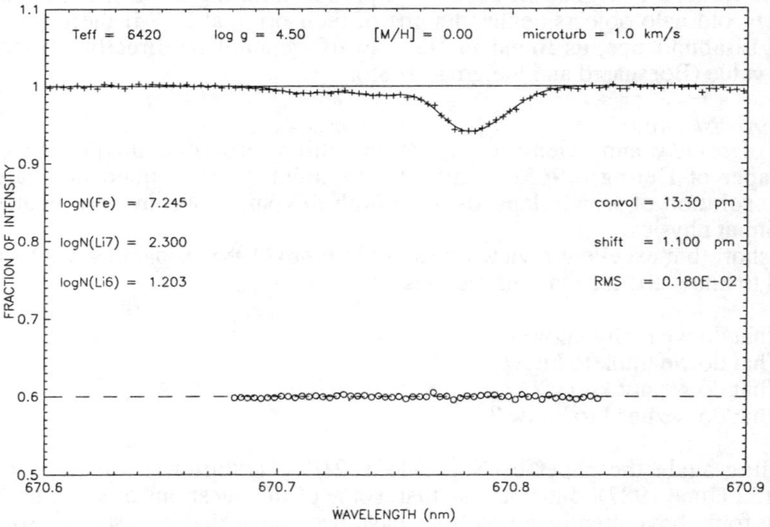

Fig. 4 b. The same as Fig.4a , except that the meteoritic ratio ^{6}Li /^{7}Li = 0.08 has been imposed. A fit of similar quality has been observed (RMS = 0.00180), but at the expense of a wavelength shift which is 4 time the RMS error on the wave length scale

"Lithium in lower main sequence stars of the α Per cluster" have found that stars belonging to the same cluster, and having the same T_{eff}, may have very different Li-abundance. Schatzman, in this Conference in order to explain high and low Li-abundance in such stars, has invoked the role of rotational braking, involving dynamo and mass loss as well as differences in accretion disks in pre-main-sequence phase.A very interesting paper on Li-depletion and rotation in main sequence stars by Balachandran (1990) will appear in the Astrophys.J..Nearly 200 F-stars have been analysed by Balachandran in order to examine the effects of rotational braking on Li-abundance.

"Is Li as rare as we think it is in stars less massive than the Sun ?" This is one of the questions in section No 4 which was preceded in the same section by a recommandation that Li surveys of F, G and K stars should be made and become available. We are pleased to announce that such surveys are afoot. Pallavicini et al. (1987) have determined Li abundances in 27 lower main sequence stars and have discussed Li in each of these stars.

As a byproduct of spectral analyses of G and K dwarfs within 10 pc of the Sun (Perrin et al. 1988) a strong lithium line has been found in the cool K2V star, HD 17925 (Cayrel de Strobel and Cayrel 1989). The Li-abundance of HD 17925, log N(Li) = 2.50 ± 0.10 (on the scale log N(H) = 12.0) is somewhat smaller than the cosmic abundance, but the rate of Li depletion in a low mass star as HD 17925, ($\mathbf{M_*} \approx 0.8\ \mathbf{M_\odot}$) is high. Therefore the presence of the Li-line indicates that HD 17925 is very young, not more than a few million years old. About Soderblom's question on the rarity of Li-rich low mass stars, we cannot answer with only one Li-rich disk-K-star found, we think, nowever that Li-rich K disk dwarfs are rare in the solar neighbourhood, which contains mostly old disk field dwarfs.

We will finish this paragraph in commenting a point of the first section of the paper by Soderblom : "What do we really know ?". We really know that the "precipitous dip among the mid-F dwarfs" which Boesgaard discovered in the Hyades (Boesgaard and Tripicco 1986a, 1986b) does not only exist in the Hyades, but also in Praesepe and Coma clusters and in U Ma and Hyades moving groups (Boesgaard and Budge 1988, Boesgaard et al. 1988). A very interesting research on Li in some subgiants of the old open cluster M67 has been done by Balachandran (1990) very recently. Apparently the"Boesgaard Li-chasm" occurs for clusters older than 3 x 10^8 yr, because the much younger α Per and Pleiades clusters do not show the Li-chasm (Boesgaard et al. 1989).
Some theoreticians tried to explain the physical processes implied by the existence of the chasm. Michaud proposes to explain this feature by selective diffusion processes.As an alternative,S.Vauclair(1988) has proposed that the chasm is produced by turbulence induced by rotation,according a hydrodynamic model due to Zahn (1990) Recently Charbonneau and Michaud (1989, 1990) found that meridional circulation possibly could be responsible for Boesgaard's chasm.

In the next paragraph we will try to find out which of these theories is the more convincing. As working tool we shall use the ^{6}Li/^{7}Li ratio of some F-stars falling in the Li-dip.

5.3 ^{6}Li/^{7}Li *ratio as a test for the physical processes responsible of the Li-chasm in open cluster F-dwarfs*

If selective diffusion is the process of the abrupt depletion of Li observed by Boesgaard and coworkers in middle F stars, ^{6}Li should be slightly less depleted than ^{7}Li in stars falling in the Li-chasm. On the contrary, in high turbulent mixing and in meridional circulation burning is the depletion agent and would affect much more strongly ^{6}Li than ^{7}Li. It is therefore quite interesting to find out what happens to this isotopic ratio across the Li-chasm.

The ^{6}Li/^{7}Li ratio has been previously determined in solar type stars (Andersen et al. 1984, Maurice et al. 1984, Hobs 1985, Rebolo et al. 1986). Cayrel and Cayrel de Strobel (1990) have attempted to measure more accurately this ratio in 14 F dwarfs using the larger light collecting power of the CFH Telescope and trying to achieve a S/N ratio of 500 with the new holographic grating giving a resolution of 0.1Å (≈ 70000). They found that in the Boesgaard chasm good fits with the meteoritic ratio : ^{6}Li/^{7}Li = 0.08 can be obtained only by shifting the expected wavelength of the Li-blend by several time the RMS error expected on wavelength determination. Applying to the nearby FeI line at 6705.11Å the shift, found for the Li line, would be unacceptable. Therefore selective diffusion does not seem favoured at this

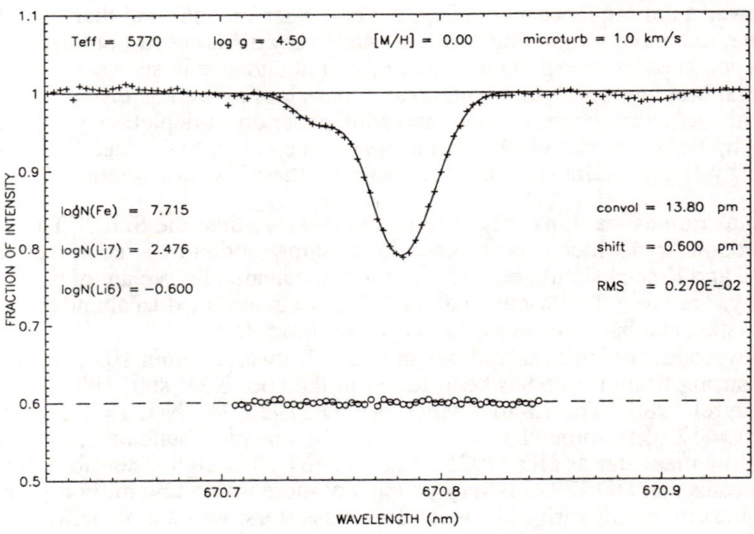

Fig.5 a. Observed and synthetic spectra of the Li-blend region for HD1835 .The Li-abundance is set to a negligeable value. The RMS value of the fit is : 0.00270. The wavelength shift applied is 0.600 pm or 6 m Å .

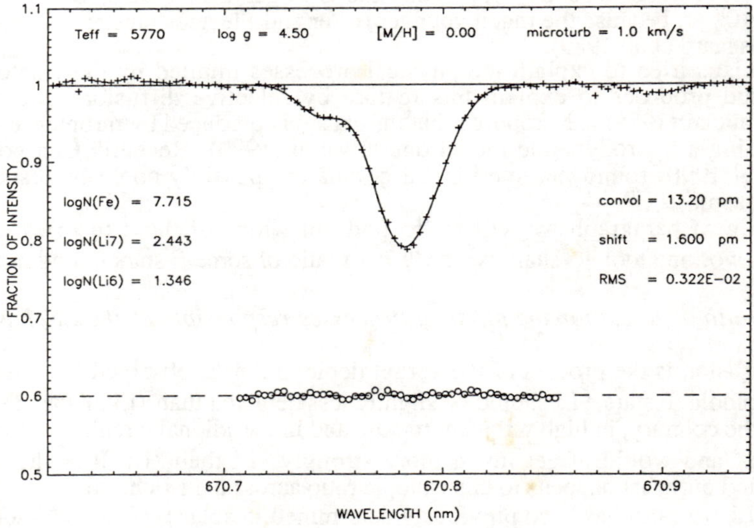

Fig.5 b The same as Fig.5 a, except that the meteoritic abundance ratio : $^{6}Li / ^{7}Li = 0.08$ has been imposed. Here both the RMS of the fit is worse, and the wavelength shift is too large.

time, because in our observations there is not a single case for which there is a positive detection of ^{6}Li. However Proffitt and Michaud (1989), have shown that ^{6}Li can be destroyed already in the pre-main-sequence. In that case, its absence in the Boesgaard chasm, does not prove anything for the Li-depletion mechanism.

In figures 4a and 4b are shown observed and synthetic spectra of the region of the Li-blend for the star HD 211976 (Teff = 6420 K) ,which falls in the Boesgaard Li-chasm. In figures 5a and 5b are shown observed and synthetic spectra of the Li-blend for HD 1835, a moving group star of the Hyades, having the same temperature than the Sun (Teff = 5770 K). We have chosen these two stars as example:they are particularly suitable because of their very low rotational broadening.

6. Conclusion

The purpose of this review was to show to non specialists what kind of physical processes are affecting surface stellar abundances. Abundance trends in pop I and pop II stars have been presented. Only abundance results coming from high S/N, high resolution solid state spectroscopy have been discussed. Many references have been given, they are almost all very recent, and their important number shows the general astronomical interest in surface elemental abundances. Nevertheless, not all of the surface abundances in stars have been discussed. Abundances in pre-main-sequence stars, in white dwarfs, in the coolest main-sequence stars, in AGB stars, in intrinsic variables, have been left out.

This paper was given and written in remembering 37 years of friendship with Evry Schatzman, and for a similar amount of time permanent struggling with stellar abundances.

References

Abbott, D.C. and Hummer D. G. : 1985, Astrophys. J. **294**, 286

Alecian, G., Artru, M.C. : 1987, Astron. Astrophys **186**, 223

Alecian, G. Artru, M.C. : 1988, IAU Symp. N°132. The impact of very high S/N spectroscopy on stellar physics, Cayrel de Strobel, G., Spite M., eds p. 235

Alecian, G. Vauclair, S. : 1983, Fundamentals of Cosmic Physics 8, 369, Gordon and Breach Science Pub.

Anders, E., Grevesse, N.: 1989, Geochimica and Cosmochimica Acta 53, n°1

Andersen, J., Gustafsson, B., Lambert, D.L. : 1984, Astron. Astrophys.**136**, 65

Baglin, A., Lebreton,Y 1990, "Inside the Sun", Symposium held in Versailles, France, Mai 1989 (in press)

Balachandran, S., Lambert, D.L., Stauffer, J.R. : 1988, Astrophys. J. **333**, 267

Balachandran, S.; 1990, to appear in Astrophys. J.

Balachandran,S. : 1990,to appear in the proceedings of the sixth Cambridge workshop on : "Cool stars , stellar systems, and the Sun "

Bessell, M.S., Norris, J. : 1987, J. Astrophys Astr. **8**, 99

Bohannan, B., Abbott, D.C., Voels, S.A., Hummer D.G. : 1986, Astrophys. J. **308**, 728

Bohannan, B., Voels, S.A., Abbott D.C., Hummer D.G. : 1988, IAU Symp. N° 132, The impact of very high S/N spectroscopy on stellar physics, G. Cayrel de Strobel, M., Spite eds. p. 127

Bohlender, D.A., Brown D.N., Landstreet, J.D., Thompson, I.B. : 1987, Astrophys. J. **323**, 325

Boesgaard, A., Steigman, G. : 1985, Ann. Rev. Astron. Astrophys. **23**, 319

Boesgaard, A.M., Tripicco, M.J. : 1986a, Astrophys. J. Letters **302**, L49

Boesgaard, A.M., Tripicco, M.J. : 1986b, Astrophys. J. **303**, 724

Boesgaard, A.M., Budge, K.G. : 1988, Astrophys. J. **332**, 1

Boesgaard, A.M., Budge, K.G., Burck, E.E. : 1988, Astrophys. J. **325**, 749

Boesgaard, A.M. : 1989, Astrophys. J. **336**, 798

Boesgaard, A.M., Budge, K.G., Ramsay, M.E. : 1989, Astrophys. J. (in press)

Burkhart ,C.,Coupry, M.F.,Van't Veer C.: 1988, IAU Symp. 132,The impact of very high S/N spectroscopy on stellar physics, G. Cayrel de Strobel M. Spite eds
Burkhart, C., Coupry, M.F. : 1989 private communication
Carney, B.W., Latham, D.W., Jones, R.V., Beck, J.A., Laird, J.B. : 1989, The abundance spread within globular clusters : spectroscopy of individual stars, G. Cayrel de Strobel, M. Spite, T. Lloyd Evans, eds (Observatoire de Paris) p. 87
Castelli, F., Faraggiana, R., Hvala, S. : 1976, Astron. Astrophys. **46**, 99
Cayrel de Strobel, G., Bentolila, C. : 1983, Astron. Astrophys. **119**, 1
Cayrel, R., Cayrel de Strobel, G., Campbell, B., Däppen, W. : 1984, Astrophys. J., **283**, 205
Cayrel de Strobel G., Bentolila, C., Hauck, B., Duquennoy A. : 1985, Astron. Astrophys. Suppl. Ser **59**, 145
Cayrel de Strobel, G. : 1985, "La composition chimique des étoiles dans le voisinage solaire", Comptes Rendus sur les journées de Strasbourg, 7ème réunion, Observatoire de Strasbourg ed., pp. 8
Cayrel, R., Cayrel de Strobel, G., Campbell, B. : 1985, Astron. Astrophys. **146**, 249
Cayrel de Strobel, G. : 1987, J. of Astrophys. and Astron. **8,** 141
Cayrel, R. : 1988, IAU Symp. n°132, The impact of very high S/N spectroscopy on stellar physics, G. Cayrel de Strobel, M.Spite eds; p.345
Cayrel, R., Cayrel de Strobel, G., Campbell, B. : 1988, IAU Symp. 132, the Impact of very high S/N Spectroscopy on stellar Physics,G. Cayrel de Strobel,M. Spite eds. p.449
Cayrel de Strobel, G., Cayrel, R. : 1989, Astron. Astrophys. **218,** L9
Cayrel de Strobel,G.,Bentolila,C.:1989, Astron. Astrophys.**211**, 324
Cayrel R., Cayrel de Strobel, G., Vauclair, S., Bentolila, C.:1990:,to appear. in Astron. Astrophys.
Charbonneau, P., Michaud, G., Proffitt, C.R. : 1989, Astrophys. J. **347**, in press
Charbonneau, P., Michaud, G. : 1990, Astrophys. J. 351, in press
Cowley C.R., Jaschek, M., Acker, A. : 1985, Astron. Astrophys. **149**, 224
Duncan, D.K., Jones, B.F. : 1983, Astrophys. J. **271**, 66
Gigas, D. : 1986, Astron. Astrophys. **165**, 172
Gigas, D. : 1988, Astron. Astrophys. **192**, 264
Gigas, D. : 1988, IAU Symp. n°132, The impact of very high S/N spectroscopy on stellar physics G.Cayrel de Strobel, M.Spite eds p. 389.
Gilmore, G., Reid, N. : 1983, Mon. Not. Roy. Astron. Soc. **207**, 223
Gilmore, G., Wyse, R.F. : 1985 Astron. J. **90,** 2015
Grenon, M. : 1987, J. Astrophys. Astron. **8,** 123
Gustafsson, B.: 1987, "Stellar evolution and dynamics of the outer Halo", M. Azzopardi and F. Matteucci, eds (ESO-Garching)
Gustafsson, B. : 1988, IAU SYM. n°132, The impact of very high S/N spectroscopy on stellar physics, Cayrel de Strobel, G., Spite,M., eds p. 333
Herbig, G.H. : 1965, Astrophys. J. **141**, 588
Hillier D.J. : 1988, Astrophys. J. **327**, 82,
Hobbs, L.M. : 1985, Astrophys. J. **290**, 284
Hobbs, L.M., Duncan, D.K. : 1987, Astrophys. J. **317**, 796
Hummer,D.,G.: 1982, Astrophys. J.,**257**,724
Kawano, L., Schramm, D., Steigman, G. : 1988, Astrophys. J. **327**, 750
Laird, J., Rupen, M., Carney, B.W., Latham,D.W./ 1989, The abundance spread within globular clusters: spectroscopy of individual stars, G; Cayrel de Strobel, M.Spite,T. Lloyd Evans, eds;(Observatoire de Paris) p.87
Lambert, D.L. : 1987, J. Astrophys. Astr. **8** , 103
Lambert, D.L. : 1988, IUA Symp. n°132, The impact of very high S/N spectroscopy on stellar physics", Cayrel de Strobel, G., Spite, M., eds. p. 563
Lambert, D.L. : 1989, Cosmic Abundances of matter, AIP Conference Proceedings 183, p. 168, Lerner, G. editor University of Minnesota
Maeder, A. : 1983, Astron. Astrophys. **120**, 113
Matteucci, F., Tornambé, A. : 1985, Astron. Astrophys. **142**, 13
Maurice, F., Spite, F., Spite, M. : 1984, Astron. Astrophys. **132**, 278

Mazzei, P., Pigatto, L. : 1989, Astron. Astrophys. **213**, L1
Megessier, C. : 1988, Astron. Astrophys. Suppl. Ser **72**, 551
Megessier, C. : 1989, private communication
Michaud, G. : 1980 Astron. J. **85**, 589
Michaud, G., Megessier, C., Charland, Y. : 1981, Astron. Astrophys. **103**, 244,
Nissen,P.E.; 1990,XI IAU European Astronomical Meeting, "New Windows on the Univers"ed. M. Vazquez, Cambridge Univ. Press
Pallavicini, R., Cerruti-Sola, M., Duncan, D.K. : 1987, Astron. Astrophys. **174**, 116
Perrin, M.N., Cayrel de Strobel, G., Dennefeld, M.: 1988, Astron. Astrophys. **191**, 237
Proffitt, C.R., Michaud, G. : 1989, Astrophys. J. **346**, 376
Przybylski, A. : 1977, Mon. Not. Roy. Astron. Soc. **178**, 71
Rebolo,R.,Crivellari,L.,Castelli,F.,Foing,B.,Beckman,J.E.:1986,Astron.Astrophys.**166**,195
Rebolo, R., Molaro, P., Beckman, J. E. : 1988, Astron. Astrophys. **192**, 192
Schatzman, E. : 1977, Astron. Astrophys. **56**, 211
Schatzman, E. : 1989a Turbulence and nonlinear dynamics in MHD flows. Eds. Meneguzzi, M., Poquet, A. Sulem, P.L., Edts., Elsevier Science publishers
Schatzman, E. : 1990, in the Proceedings of this Colloquium
Soderblom, D.R. : 1988, IAU Symp. N.132, "The impact of very high S/N Spectroscopy on stellar physics", G. Cayrel de Strobel and M. Spite eds., p. 381
Spinrad, H., Taylor, B.J. : 1969, Astrophys. J. **157**, 279
Spite, M., Spite, F. : 1982a, Nature 297, 483
Spite, F., Spite, M. : 1982b, Astron. Astrophys. **115**, 357
Spite,M.,Maillard,J.P.,Spite ,F.: 1984,Astron. Astrophys;**141**,56
Spite, M., Spite, F. : 1985, Ann. Rev. Astron. Astrophys, **23**, 225
Spite F., Spite, M.:1986 Astron. Astrophys. **163**,140
Spite, M.: 1990, to appear in the Proceedings of the "Elba Workshop 1989"
Twarog, B.A, : 1980, Astrophys. J. **242**, 242
Van't Veer,C., Faraggiana,R.,Gerbaldi,M.,Castelli,F., Burkhart,C.,Floquet,M.:1990, Astron. Astrophys.,. in press
Vauclair,G.,Vauclair,S.:1978,Astrophys. J.**223**,920
Vauclair, S. : 1988, Astrophys. J. **335**, 971
Voels, S.A., Bohannan, B., Abbott, D.C., Hummer D.G. : 1989, **340**, 1073
Wagoner, R.V. : 1973; Astrophys. J. **179**, 343
Wegner, G., Cummins, D.J., Byrne, P.B., Stickland, D., J. : 1983, Astrophys. J. **272**, 646
Wheeler, J.C., Sneden, C., Truran, J.W. : 1989, Ann. Rev. Astron. Astrophys. **27**, 279
Withford, A.E., Rich, R.M. : 1983, Astrophys. J. 274, 723
Wolff, S.C., Hagen, W. : 1976, Publ. Astron. Soc. Pacific **88**, 119
Wolff, S.C., Heasly, J.N. : 1985, **292**, 589
Wyse, R.F., Gilmore, G. : 1986, Astron. J. **91**, 855
Yang,J.,Turner,M.S.M.S.,Steigman,G.,Schramm,D.H,Olive,K.A.:1984,Astrophys.J., **281,** 493
Zahn,J.P.:1987a, " The internal solar angular velocity "eds. B.R.Durney and S.Sofia Dordrecht p.201
Zahn,J.P.:1987$_b$, "Astrophysical fluid dynamics", (Les Houches),eds.J.P.Zahn and J.Zinn Justin,New Jork: Elsevier Science Publishers
Zahn, J.P. : 1990 , in the Proceedings of this Colloquium

STELLAR ROTATION PRIOR TO THE MAIN SEQUENCE

Jérôme Bouvier
Canada-France-Hawaii Telescope Corporation
P.O. Box 1597, Kamuela, HI 96743

I. Introduction

While large–scale studies of rotation in massive main sequence stars began more than 30 years ago, very little was known about the rotational velocities of pre–main sequence stars prior to the early 80's. Since then, the advent of more sensitive detectors and the development of improved spectral analysis techniques made it possible to derive the rotational velocities of large samples of pre–main sequence stars. Two complementary observing methods were used. One consists in measuring the Doppler broadening of line profiles, which yields an estimate of the star's rotational velocity projected onto the line of sight, $v\sin i$. The relative faintness of solar–mass pre–main sequence stars often prevents observations at very high spectral resolution, so that $v\sin i$'s less than 10 km s^{-1} cannot be usually measured. The other method applies to stars which exhibit temperature inhomogeneities at their surface, i.e., "spots". As spots are carried across the stellar disk by rotation, the star's luminosity is modulated with a periodicity which directly reflects its rotational period. Since rotational modulation is free of projection effects, it provides more useful estimates of the star's rotation rate than spectroscopic methods. The application of both methods to pre–main sequence stars has now provided a large enough database to characterize their rotational properties on a statistical basis.

Pre–main sequence stars are the first visible manifestation of newly–formed objects. Therefore, their rotational properties are expected to bear information relevant to the evolution of angular momentum during the star formation process. In contrast, by the time solar–mass stars are on the main sequence, any spread in the initial angular momentum distribution has been smoothed out, or is hidden in the star's interior, as indicated by their uniformly low surface rotation rates. As stars age, rotational velocity variations are caused both by structural changes in the stellar interior and by the star's interaction with its environment. During pre–main sequence evolution, dramatic alteration of the internal structure of low–mass stars occurs as a radiative core develops at the center of the initially completely convective star. Moreover, recent models suggest that young stars strongly interact with their surroundings both by driving powerful winds and by accreting from a circumstellar disk. Therefore, the study of rotation in pre–main sequence stars is expected to bring deep insight into the physical mechanisms which control angular momentum transfer during the star's lifetime.

Recent reviews of rotation in low–mass stars, from pre– to post–main sequence phases of evolution, have been published by Stauffer and Hartmann (1986) and Stauffer and Soderblom (1989). In the present review, the evolution of angular momentum is described from the pre–stellar dark cloud stage to the zero–age main sequence residence.

In Section II, the evolution of angular momentum during the star formation process is discussed in the light of recent measurements of rotation in dark clouds and in low–mass pre–main sequence stars. In Section III, the rotational properties of pre–main sequence stars with masses from less than 0.5 up to 5 $M_\odot$ are reviewed. Emphasis is placed on the variation of mean rotational velocity with mass and on the comparison between the rotational properties of weak and strong emission–line stars. The evolution of angular momentum as the star contracts toward the main sequence is investigated in Section IV.

II. Angular momentum evolution during star formation

Measurements of rotation in dark clouds and in pre–main sequence stars bring clues about the evolution of angular momentum during the star formation process. Studies of angular rotation in dark clouds shed light onto their angular momentum distribution and allow one to follow their rotational evolution as they slowly contract before the protostellar collapse is initiated. Unfortunately, the evolution of angular momentum during the protostellar gravitational collapse cannot be traced observationally because protostars are deeply embedded in dusty cocoons which are opaque to visible light. Hence, the next piece of information astronomers have on the early evolution of stellar angular momentum is provided by the rotational velocities of newly–formed, visible pre–main sequence stars. The observed distribution of rotational velocities in pre–main sequence stars provides strong constraints onto the mechanisms which are responsible for angular momentum loss during star formation. In this section, a brief account of recent observational results and current models is given.

a. The angular momentum problem

The essence of the angular momentum problem can be stated in the classical following way. From interstellar matter to stars, the mean density increases by 24 orders of magnitude. Conservation of angular momentum ($J = k \cdot M \cdot R^2 \cdot \Omega$, where k is a constant of the order of unity, M the mass, R the radius, and Ω the angular velocity) implies that the angular velocity increases with mean density as:

$$\Omega/\Omega_o = (n/n_o)^{2/3}$$

A lower limit to Ω_o may be taken as the mean galactic rotation, $\Omega_{gal} \simeq 10^{-15}$ s^{-1}, and $n_o \simeq 1$ cm^{-3} for diffuse interstellar gas. At stellar densities ($n \simeq 10^{24}$ cm^{-3}), the predicted angular velocity is several orders of magnitude larger than the critical velocity at which centrifugal force balances gravity at the stellar equator. Thus, assuming angular momentum conservation and provided that the above estimates are correct to at least orders of magnitude, no star would form.

Measurement of rotation rates in dark clouds and in pre–main sequence stars allows one to state the angular momentum problem in more precise terms. The results of several studies aimed at measuring the rotation rates of pre–main sequence stars with a mass less than 3 $M_\odot$ (T Tauri stars) are illustrated in Figure 1. In this figure, all the T Tauri stars with known rotational velocities are plotted in a theoretical H–R diagram. Open and dark circles represent weak and strong emission–line stars, respectively, and

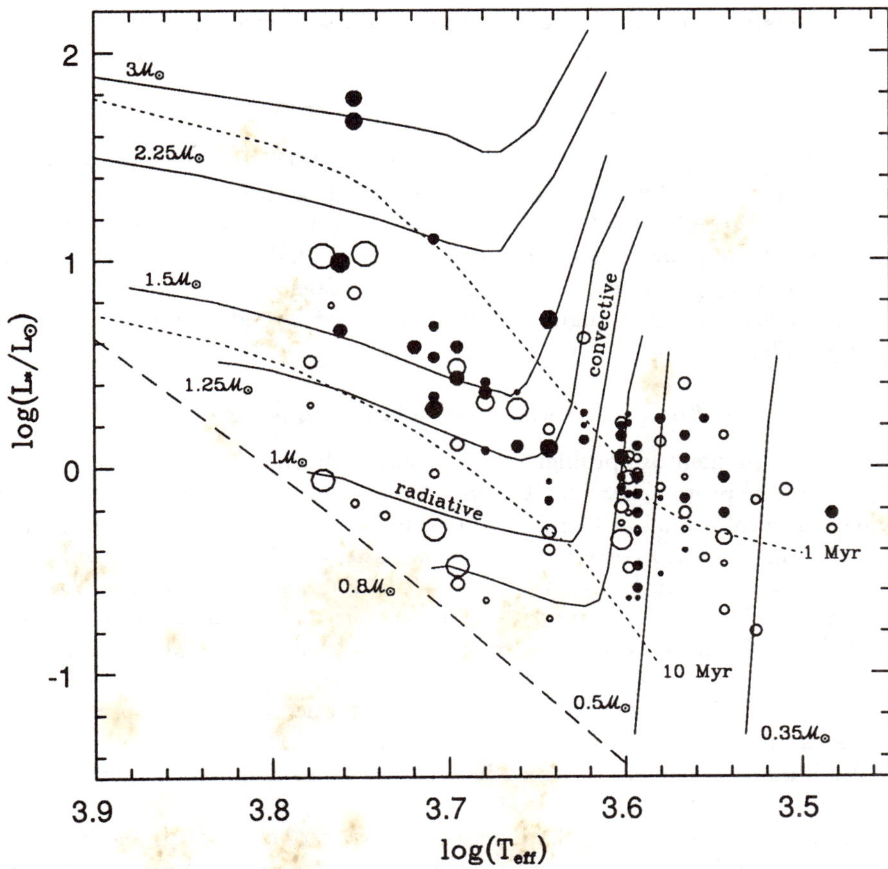

Figure 1. Rotational velocities of low-mass pre-main sequence stars in the H-R diagram. Open and dark circles represent T Tauri stars with H_α equivalent widths smaller and greater than 10 Å, respectively, and the circle area is proportional to $v\sin i$ in the range from less than 10 to about 100 km s^{-1}. *Solid lines:* theoretical pre-main sequence evolutionary tracks for stars in the mass range from 0.35 to 3 $M_\odot$ (from Cohen and Kuhi 1979). *Dotted lines:* isochrones corresponding to an age of 10^6 and 10^7 years, respectively. *Dashed line:* theoretical zero-age main sequence.

the circle area is proportional to the stellar velocity projected on the line of sight, $v\sin i$ (references for the rotational data may be found in Bouvier 1990). The rotational velocities of T Tauri stars range from less than 10 up to 100 km s^{-1}, with a mean $v\sin i$ of the order of 25 km s^{-1}, i.e., a factor of 10 below break-up velocity ($v_b = (GM_\star/R_\star)^{1/2} \simeq 300$ km s^{-1} for a solar-mass T Tauri star). This indicates that the bulk of the angular momentum problem is solved before stars appear as visible objects (Vogel and Kuhi 1981).

For a completely convective, solar–mass pre–main sequence star with $R_\star = 2\ R_\odot$ and a mean rotational velocity $v_{rot} = 25$ km s^{-1}, the specific angular momentum is:

$$(J/M)_\star = 0.2 \cdot v_{rot} \cdot R_\star = 7.10^{16} cm^2 s^{-1}$$

assuming solid–body rotation. This is to be compared with the specific angular momentum initially contained in diffuse molecular clouds from which stars ultimately form. Measurements of rotation in dark clouds indicate that low-density clouds ($n = 10^2$ cm^{-3}) have a specific angular momentum of the order of 10^{24} cm^2 s^{-1} (Goldsmith and Arquilla 1985). Hence, going from the diffuse cloud stage to the first appearance of stars at visible wavelengths, the angular momentum must be reduced by roughly seven orders of magnitude.

b. The quasi–static cloud contraction phase

Part of the angular momentum problem may be solved during the early stages of cloud contraction before the protostellar gravitational collapse begins. Goldsmith and Arquilla (1985) reviewed the rotational properties of molecular clouds with densities in the range from 10^2 to 10^4 cm^{-3}, which corresponds to cloud sizes of 10 and 0.1 pc, respectively. They find a tight correlation between specific angular momentum and cloud size and derive the relationship: $J/M \sim (size)^{1.4}$. The specific angular momentum decreases from 10^{24} cm^2 s^{-1} for massive clouds ($n = 10^2$ cm^{-3}, $M = 10^3\ M_\odot$), to 10^{21} cm^2 s^{-1} for dense molecular cloud cores of a few solar masses ($n = 10^4$ cm^{-3}). Although Goldsmith and Arquilla's study is biased toward the detection of rapidly–rotating clouds (see Shu, Adams, and Lizano 1987), their results suggest that, if molecular cloud cores are the evolutionary descendents of massive clouds, large angular momentum losses occur during this phase of cloud evolution.

Bodenheimer (1978) suggested that successive phases of cloud fragmentation could efficiently convert angular momentum of spin into orbital motion. Alternatively, magnetic braking may be instrumental in rotating clouds during the quasi–static contraction phase (Mestel and Spitzer 1956). Mouschovias and Paleolougo (1979, 1980) worked out in some details the influence of microgauss interstellar fields upon the dynamics of diffuse clouds as they contract to higher densities. For typical physical properties of relatively diffuse clouds ($n = 10^3$ cm^{-3}, $M = 10^3\ M_\odot$), they find that the ionization fraction is high enough for the neutrals to be efficiently coupled to the magnetic field via neutrals–ions collisions. As a result, angular momentum is redistributed by coupling the rotational motion of the cloud to the slowly–rotating outer envelope and the cloud contracts at nearly constant angular velocity. They claim that this process can account for a reduction of the angular momentum by a few orders of magnitude by the time densities typical of molecular cloud cores ($n = 10^{4-5}$ cm^{-3}) are reached.

That magnetic fields play an important role in the dynamics of molecular clouds is illustrated by Heyer *et al.*'s (1987) study of 5 dark clouds in the Taurus region. These clouds, with a mass of several $10^2\ M_\odot$ and densities between 10^3 and 5.10^3 cm^{-3}, exhibit an elongated shape with the minor axis parallel to the local direction of the interstellar magnetic field. This suggests that cloud contraction has preferentially occurred along magnetic field lines as lateral contraction is inhibited by magnetic pressure. Moreover, they found the rotational axes of the clouds to be aligned with the direction of the ambient magnetic field as expected from magnetic braking processes.

During the diffuse stage, quasi–static cloud contraction in the direction perpendicular to the field lines occurs due to the slow drift of the neutral material through both the ions and the magnetic field. This process is responsible for magnetic flux leakage and is often referred to as ambipolar diffusion (see e.g., Spitzer 1978). As the cloud density increases, the ionization fraction decreases and neutrals–ions coupling becomes less efficient. The time scale for ambipolar diffusion shortens and, above a critical density, the rapidly decreasing magnetic flux becomes too weak to support the gas against gravity. At this point gravitational collapse takes place at free–fall rate and angular momentum is subsequently conserved.

The amount of angular momentum lost during the slow contraction phase depends crucially upon the value of the critical density at which the magnetized cloud becomes unstable to dynamical collapse. Theoretical estimates range from 10^5 to 10^9 cm^{-3} (Mouschovias 1977, Nakano 1979), while observations suggest somewhat lower values. Heyer (1988) showed that the rotational axes of 13 molecular cloud cores with $n \simeq 0.5-2\ 10^4$ cm^{-3} in the Taurus region were randomly distributed relative to the direction of the ambient magnetic field. This stands in sharp contrast with the general alignment of the rotational axes of more diffuse clouds ($n \simeq 10^3$ cm^{-3}, Heyer *et al.* 1987) with the local magnetic field direction and suggests that magnetic field ceases to be dynamically important at densities typical of molecular cloud cores ($n \simeq 10^4$ cm^{-3}). Heyer (1988) proposes that randomization of the rotational axes occurs at this point via gravitational interaction or physical collisions between cores.

If magnetic coupling remains effective up to densities larger than $\simeq 10^7$ cm^{-3}, rotational braking may entirely solve the angular momentum problem (e.g., Norman and Heyvaerts 1985, Mouschovias and Paleolougo 1986). If, however, magnetic support becomes inefficient at densities typical of molecular cloud cores as suggested by observations, the angular momentum at the start of the protostellar collapse still greatly exceeds that of young solar–mass stars (10^{21} and 10^{17} $\mathrm{cm}^2\ \mathrm{s}^{-1}$, respectively). An additional mechanism must then be instrumental in removing the angular momentum in excess during the protostellar collapse.

c. *Protostellar collapse*

Unfortunately, because of the large obscuration which characterizes collapsing clouds, there is an observational gap between molecular cloud cores and pre–main sequence stars. Therefore, one must rely on theoretical models in order to investigate the fate of the angular momentum during the protostellar collapse.

Terebey, Shu, and Cassen (1984) investigated the gravitational collapse of the inner 1 $M_\odot$ part of a molecular cloud core assumed to be an isothermal sphere in a state of uniform rotation with $\Omega = 10^{-14}$ s^{-1}. Assuming angular momentum conservation, they find that about half of the collapsing material falls directly to the center to form a rapidly–rotating protostar while the rest accumulates in a protostellar disk which contains a large fraction of the initial angular momentum. Alternatively, Bodenheimer (1978) suggested that fragmentation during gravitational collapse efficiently converts spin into orbital momentum. Boss (1986) performed 3–D collapse calculations starting from uniform density clouds and showed that, for initial conditions typical of molecular clouds cores, the output of gravitational collapse includes both single and binary protostars having specific angular momenta of the order of 10^{18-19} cm^2 s^{-1}. Thus, while both fragmentation processes and the formation of a protostellar disk may substantially alleviate the angular momentum problem, newly–formed protostars are still predicted to rotate close to break–up velocity, i.e., much faster than visible pre–main sequence stars.

Further reduction of the angular momentum of protostars may result from the onset of strong winds. Observations of stellar formation regions at millimeter wavelengths reveal the existence of large–scale (a few 0.1 pc) molecular CO outflows usually associated with deeply embedded young stellar objects. Molecular outflows are thought to result from the interaction of an energetic protostellar wind with the surrounding molecular environment and such episodes of strong mass ejection are now believed to be a universal phase of early stellar evolution (see a review by Lada 1985). Bertout, Basri, and Cabrit (1990) reviewed the various wind models proposed so far to account for the large momentum rate and mechanical luminosity observed for CO outflows. Although they find none to be entirely satisfactory, they argue that the most promising mechanisms are those which rely upon the combination of fast rotation and magnetic field, either in the protostar or in the protostellar disk. A common feature of these models is that they predict large angular momentum loss on a time scale of the order of 10^{4-5} years (e.g., Hartmann and McGregor 1982, Draine 1983, Pudritz and Norman 1986). Therefore, even though the lifetime of CO outflows is fairly short (a few 10^5 years), protostellar rotation may be strongly braked by a magnetized wind before stars appear at optical wavelengths. This mechanism may in fact explain why young visible stars have rotational velocities far below breakup, even though, in order to form a star, there is no need to remove more angular momentum than what would leave it at break–up. The main uncertainty here lies in the unknown origin and strength of protostellar magnetic fields which primarily control the efficiency of the magnetic braking process.

d. Conclusion

Despite theoretical and observational uncertainties associated with each stage of the star formation process, substantial advances have been made to account for the large angular momentum losses which characterize the earliest phases of stellar evolution. While current models seem to be able to handle the angular momentum problem, they suggest that the early evolution of angular momentum is very sensitive to the detailed physical conditions which prevail in dark clouds. For instance, the critical density at which a cloud becomes unstable to gravitational collapse may vary from one cloud to another, depending upon environmental conditions such as the proximity of ionizing sources. Consequently, clouds on the verge of collapse may exhibit quite a wide range of rotational states. This issue is of particular importance because collapse calculations indicate that the rotational properties of newly–formed stars are determined to a large extend by the ratio of rotational to gravitational energy at the start of the gravitational collapse. A wide spectrum of initial conditions for collapsing clouds may then result in a significant spread in angular momentum among pre–main sequence stars. This and other issues related to the rotational properties of pre–main sequence stars are discussed in the next section.

III. Rotation in pre–main sequence stars

The evolutionary state at which a star first becomes visible depends upon its mass. Low–mass stars appear at optical wavelengths long before they reach the main sequence (see Fig. 1), while high–mass stars evolve much more rapidly and are already near or on the main sequence when their surroundings become transparent to visible light. Hence, pre–main sequence stars with a mass less than a few solar masses are the least evolved objects whose rotational velocities can be measured.

At an age between 10^6 and 10^7 years, these stars still lie near their birthplace within large dark clouds distributed along the plane of the Galaxy. Historically, they have been divided in two broad classes according to their mass. Pre–main sequence stars more massive than 2.5 $M_\odot$ were first defined as a class by Herbig (1960) and are now known as Ae–Be Herbig stars, indicating A and B spectral types and emission features in their optical spectrum. Less massive ones, with spectral types late–F or later, are referred to as T Tauri stars, after the prototype of the class, and were first recognized as a class by Joy (1942, 1945) on the basis of their strong H_α line emission ($EW(H_\alpha) > 10$ Å and up to more than 100 Å). For many years, H_α prism objective surveys of nearby dark clouds have been the most powerful way to discover new T Tauri stars. Recently, however, pre–main sequence stars exhibiting weak H_α emission ($EW(H_\alpha) \leq 10$ Å) were discovered during X–ray or CaII H and K line surveys of stellar formation regions (Walter 1986, Herbig, Vrba, and Rydgren 1986). These stars lack all the exotic properties of Joy's "classical" T Tauri stars (CTTS) and are now commonly referred to as "weak–line" T Tauri stars (WTTS, Herbig and Bell 1988).

The spectral peculiarities of CTTS include strong Balmer and metallic emission lines in the optical, large continuous energy excesses compared to standard stars at both UV and near–IR wavelengths, and an often weak photospheric spectrum due to spectral veiling, i.e., partial or total filling–in of the photospheric absorption features by emission. In contrast, the photospheric spectrum of WTTS is very similar to that of late–type dwarfs, with only weak H_α and CaII H–K emission visible at low spectral resolution, and, while some WTTS exhibit small continuous energy excesses at near–IR wavelengths, none does in the UV domain.

Current models for WTTS and CTTS have been recently reviewed by Bertout (1989). Several lines of evidence suggest that many young stars are surrounded by circumstellar disks and it is now widely believed that the exotic properties of CTTS primarily result from the interaction between the central star and its disk. Bertout, Basri, and Bouvier (1988) showed that the UV and near–IR continuous excesses of several CTTS can be accounted for by assuming accretion of matter from a Keplerian disk onto the stellar surface at a rate ranging from 5.10^{-8} to 5.10^{-7} $M_\odot$ yr^{-1}. Basri and Bertout (1989) further suggested that the star/disk interaction may significantly contribute to CTTS radiative losses both in the Balmer jump and in the higher lines of the Balmer series. While CTTS accrete material from their circumstellar disk at appreciable rates, interpretation of the strong and broad (a few hundred km s^{-1}) H_α line profile they exhibit suggests that they simultaneously drive strong winds, with mass–loss rates in the range from approximately 10^{-8} to a few 10^{-7} $M_\odot$ yr^{-1} (see Catala, this volume).

The absence of detectable UV excess and of spectral veiling in WTTS indicates that they do not accrete material from a circumstellar disk at appreciable rates ($\dot{M}_{acc} \leq 10^{-8}$ $M_\odot$ yr^{-1}, see Bertout, Basri, and Bouvier 1988). Also, WTTS show small or no near-IR excess, which indicates a relative paucity of warm dust very near the star. Indeed, the spectral properties of WTTS from X-ray to near-IR wavelengths can usually be understood as the result of solar-type chromospheric and coronal activity driven by a magnetic dynamo (Walter *et al.* 1988, Bouvier 1990). Nevertheless, a recent re-examination of IR colors of WTTS by Strom *et al.* (1989b) revealed that a number of these stars, while exhibiting small near-IR excess, have significant mid- and far-IR excesses, indicative of cool dust located at large distances from the stellar surface. They proposed that these stars are surrounded by disks whose inner regions are relatively devoided of matter, and thus do not strongly interact with the star.

The rotational properties of Herbig Ae-Be and T Tauri stars are summarized below. The run of mean rotational velocity with mass is investigated and the rotational properties of WTTS and CTTS are compared.

a. Mass-dependence

The dependence of *v*sin*i* upon mass for pre-main sequence stars is shown in Figure 2, where TTS younger than 10^7 years and Herbig Ae-Be stars are plotted. The mass of pre-main sequence stars was derived from the comparison of the star's location in the H-R diagram with theoretical convective-radiative evolutionary tracks as illustrated in Fig. 1 and is probably accurate to within 30%. Part of the large spread observed in *v*sin*i* at any mass must reflect an intrinsic spread in equatorial velocities since neither measurement errors nor random orientation of the rotational axes† can entirely account for it (Hartmann *et al.* 1986). An intrinsic spread in the angular momentum distribution of pre-main sequence stars can be easily understood owing to the complex pre-stellar angular momentum evolution discussed in the previous section.

The most obvious feature of Figure 2 is that Herbig Ae-Be stars , with *v*sin*i* in the range from 100 to 225 km s^{-1}, rotate much faster than TTS (Finkenzeller 1984). At an age of 10^6 years, Be stars already lie onto the main sequence and Ae stars will soon reach it as late B-type dwarfs. Indeed, the *v*sin*i* distribution of Herbig Ae-Be stars is very similar to that of B0-B9 dwarfs as given by Zorec (1986)‡.

While less pronounced, an increase of rotation with mass also appears within the T Tauri class. Although they could not measured *v*sin*i* less than 25 km s^{-1} because of instrumental limitations, Vogel and Kuhi (1981) first showed that TTS more massive than 1.5 $M_\odot$ usually have larger rotational velocities than less massive TTS. Bouvier *et al.* (1986) measured *v*sin*i* for a large sample of TTS with a detection limit of a few km

---

† Weaver (1987) claimed that he finds evidence for a non-random orientation of TTS rotational axes in the Taurus region. However, as discussed by Hartmann and Stauffer (1989), the values of the stellar radii used in Weaver's study are likely to be systematically overestimated. More reliable estimates of stellar radii yield results consistent with a random distribution of TTS axial inclinations (e.g., Bertout, Basri, and Bouvier 1988).

‡ The excess of slow rotators among B0-B9 stars in the *v*sin*i* distribution given by Wolff, Edwards, and Preston (1982), and used as a comparison by Finkenzeller (1984), is due to the inclusion of stars with luminosity class III-IV (Zorec 1986).

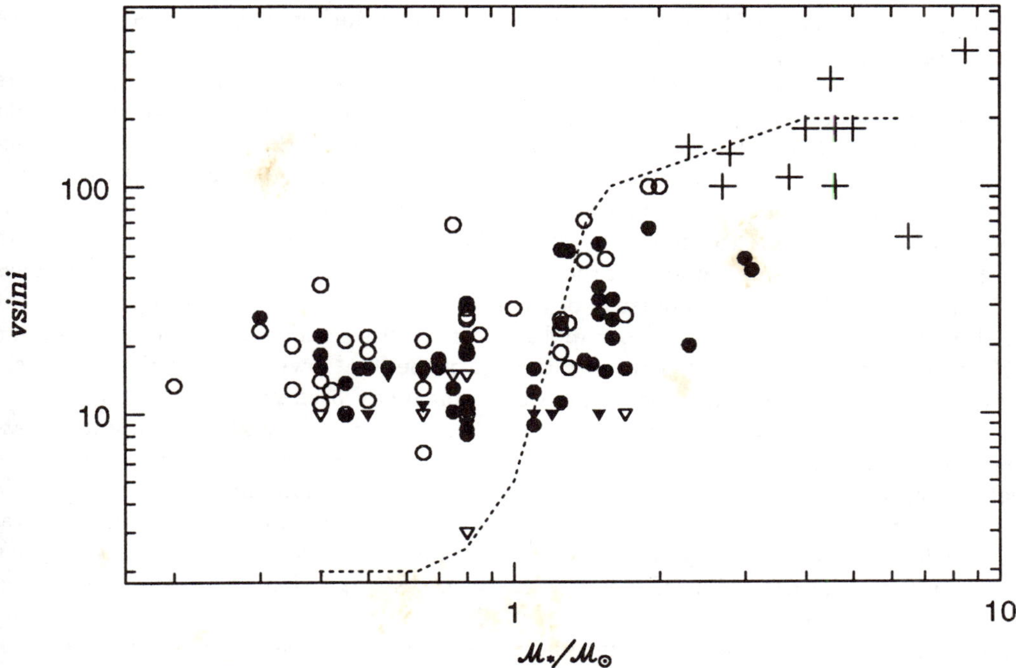

Figure 2. Rotational velocities of pre–main sequence stars younger than 10^7 years as a function of stellar mass. WTTS are represented by open circles and CTTS by dark circles, while Herbig Ae–Be stars are shown as crosses. The location of inverted triangles indicates *vsini* upper limits. The approximate run of mean *vsini* with mass observed for dwarfs is drawn as a dotted line.

s^{-1} and proposed the 1.25 $M_{\odot}$ evolutionary track as a convenient dividing line between two rotational regimes. TTS more massive than 1.25 $M_{\odot}$ have a mean *vsini* of 30 km s^{-1} while less massive ones rotate at 15 km s^{-1}, on the average. While this difference is highly significant, the *vsini* distributions of TTS on each side of the dividing line largely overlap and a smooth variation of the mean *vsini* with mass cannot be excluded from the data at hand.

As illustrated on Figure 2, a sudden drop of rotation is observed on the main sequence at spectral type F0–F4 ($M = 1.2 - 1.4$ $M_{\odot}$, Kraft 1967): dwarfs with a spectral type later than G0 usually rotate at less than 5 km s^{-1} while early–type dwarfs have a mean *vsini* of 100 km s^{-1} or more. The low rotation rates of late–type dwarfs were first interpreted by Schatzman (1962) as resulting from strong rotational braking due to magnetically–channeled mass–loss. The similar location of the rotational dividing line in TTS and in dwarfs suggests that low–mass pre–main sequence stars, which spend more time than massive ones on convective tracks, already have experienced significant rotational braking (Vogel and Kuhi 1981). There is, however, no observational evidence for a decrease of *vsini* with age along pre–main sequence convective tracks (see Fig. 1). Admittedly, derivation of TTS luminosities is difficult and makes uncertain the determination of even relative ages for these stars. More significantly, comparison between the rotational velocities of TTS and of cool dwarfs in young clusters suggests that at least some low–mass TTS are not significantly braked during their pre–main sequence evolution (see below, Section IV).

Alternatively, the observed increase of *vsini* with mass in pre–main sequence stars may be the direct output of the star formation process, i.e., more massive stars would form with higher angular momentum. Kraft (1970) found that, in early–type main sequence stars, the specific angular momentum scales with mass as: $J/M \sim M^{0.6}$. Since the rotational history of massive stars is not affected by magnetic braking, this relationship is thought to reflect the angular momentum distribution these stars had at the end of the star formation process. Assuming solid–body rotation, solar–mass TTS have specific angular momenta of the order of 10^{17} cm^2 s^{-1}, in rough agreement with what is expected from the extrapolation of Kraft's relationship to low–mass stars (Hartmann *et al.* 1986). In contrast, by the age of the Sun, low–mass dwarfs have orders of magnitude lower angular momentum due to magnetic braking.

The work of Durisen *et al.* (1986) who investigated the stability of rapidly rotating polytropes against fission may be relevant here. They showed that, for ratios of rotational to gravitational energy ($\beta = \Omega^2/4\pi G\rho$) larger than the dynamic bar mode stability limit ($\beta_{crit} \simeq 0.27$), fission instability leads to mass–loss in the form of spiral arms which may carry away up to 90% of the initial angular momentum. It is easy to show from the definition of β and J/M ($= k \cdot R^2 \cdot \Omega$), that the stability criterion ($\beta \leq \beta_{crit}$) leads naturally to $J/M \sim M^{2/3}$. Thus, the onset of fission instability during the star formation process may be ultimately responsible for the observed dependence of rotation upon mass in pre–main sequence stars (see, however, Trimble 1984 and Brosche 1986 for a critical interpretation of Kraft's relationship).

As a final remark, it should be noted that, while the rotational dividing line is located in the same mass range for pre–main sequence and main sequence stars, the amplitude of the drop in rotation between high– and low–mass stars is much smaller in TTS than in dwarfs: high–mass TTS rotate more slowly than A–F dwarfs while low–mass TTS have much higher *vsini* than late–type dwarfs. This points to quite a different rotational evolution to the main sequence for stars on each side of the dividing line (see Section IV).

b. Rotation in weak–line and classical T Tauri stars

Figure 3 shows the *vsini* distribution of WTTS and CTTS younger than 10^7 years in two mass bins. Although rapid rotators seems to preferably occur among WTTS, statistical tests indicate no significant difference between the *vsini* distributions of WTTS and CTTS (Hartmann, Soderblom, and Stauffer 1987, Walter *et al.* 1988, Hartmann and Stauffer 1989). This is somewhat surprising since, in the absence of angular momentum loss, CTTS would be expected to rotate significantly faster than WTTS as a result of accretion of high angular momentum material from their disk. Hartmann and Stauffer (1989) showed that accretion from a Keplerian disk at a rate of 10^{-7} $M_\odot$ yr^{-1} during 10^6 years would spin up a rigidly rotating, fully convective solar–mass star to about half of its break–up velocity, i.e., roughly 150 km s^{-1}. Since the observed rotation rates of CTTS are much lower, they conclude that angular momentum loss must occur and suggest rotational braking by a magnetized stellar wind as the most likely mechanism. Assuming a mass–loss rate of 3.10^{-8} $M_\odot$ yr^{-1}, they find that surface magnetic fields of a few times 10^3 gauss are required in order for the braking mechanism to balance

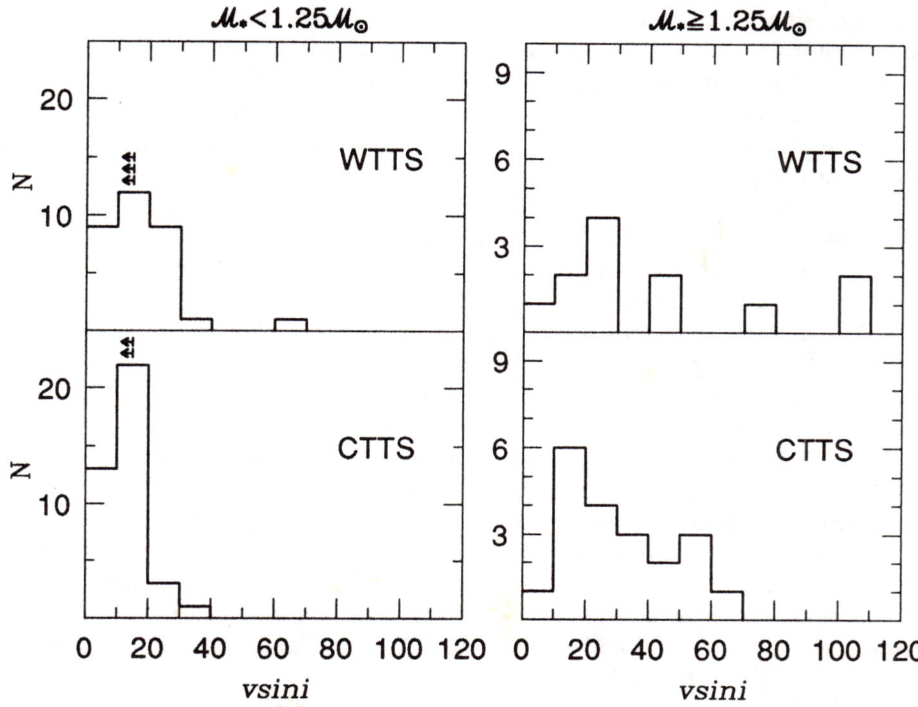

Figure 3. Distributions of *vsini* for WTTS and CTTS younger than 10^7 years. The *vsini* distributions of WTTS and CTTS with a mass less than 1.25 $M_\odot$ are shown in the left upper and lower panels, respectively, while the *vsini* distributions of more massive WTTS and CTTS are displayed in the right upper and lower panels, respectively. Stars with *vsini* upper limits of 10 km s^{-1} are included in the 0–10 km s^{-1} bin, while stars with *vsini* upper limits of 15 km s^{-1} are shown by horizontal arrows.

the angular momentum excess due to disk accretion. Although there is considerable uncertainty on this result, due to the difficulty of estimating both mass–accretion and mass–loss rates in CTTS and because of the unknown geometry of the stellar magnetic field, surface fields of a few times 10^3 gauss, i.e., similar to those observed in active late–type dwarfs, seem plausible in TTS (Bouvier 1990).

Strong rotational braking seems to be required to explain why CTTS have rotational velocities well below break–up despite disk accretion. In order to account for the similar rotation rates of CTTS and WTTS, one must further assume that angular momentum losses in the wind adjust in such a way that they exactly balance angular momentum gains from disk accretion. Cabrit *et al.* (1990) recently found a correlation between the amplitude of IR–excess and the strength of both forbidden ([OI]) and permitted (H_α) lines in TTS. Since IR excess arises in the disk while the lines they studied are formed in the wind, they interpreted this result as indicative of a causal link between accretion and mass–loss processes in CTTS. However, a proportionality between mass–loss rate and mass–accretion rate in CTTS does not imply comparable angular momentum losses and gains. Rotational braking is believed to primarily depend upon the magnetic field strength near the stellar surface, while accretion spin–up ultimately results from dissipative processes in the disk. Since angular momentum losses and gains seem to be controlled by largely independent physical mechanisms, a perfect balance between wind braking and accretion spin–up would appear rather coincidental.

A more satisfactory explanation for the similar *vsini* distributions of WTTS and CTTS may be that WTTS shared with CTTS a common rotational history, i.e., that WTTS only recently ceased to accrete from a circumstellar disk. As noted above, the hypothesis that WTTS may have evolved from CTTS as clearing of the inner disk has resulted in the termination of disk accretion is supported by the mid– and far–IR colors of WTTS reported by Strom *et al.* (1989b). Because a few WTTS seem to be younger than 3.10^6 years while some CTTS are still found at an age of 10^7 years, Strom *et al.* proposed that disk lifetimes vary from star to star within this range. Alternatively, disk accretion at rates typical of CTTS might be an episodic rather than a steady–state phenomenon, with stars on their convective tracks switching several times between quiescent (WTTS) and active (CTTS) states as a result of, e.g., dynamical instabilities in the outer disk regions. This assumption would naturally explain the similar location of WTTS and CTTS in the H–R diagram as well as their comparable rotation rates, and could also possibly account for the large overlap observed in the lithium abundances of CTTS and WTTS (Strom *et al.* 1989a, Basri, Bertout, and Martin 1990). Moreover, steady–state accretion at a rate of the order of 10^{-7} $M_\odot$ yr^{-1} and lasting 10^7 years implies a mass reservoir of at least 1 $M_\odot$ in the disk, while estimates of disk masses range from less than 0.01 to 0.1 $M_\odot$ (e.g., Chini 1989). This discrepancy can easily be solved by relaxing the hypothesis of steady–state accretion.

Finally, it should be noted that disk accretion onto magnetized objects does not necessarily lead to stellar spin–up. The torque exerted on the star by the disk consists in two parts (see, e.g., Ghosh and Lamb 1979): i) a torque due to the disk material falling onto the star which always tends to spin it up, and ii) a magnetic torque which results from the interaction of the disk material with the star's magnetic field. The magnetic torque on the central star can be either positive or negative depending upon the Alfvén radius being located inside or outside the corotation radius. For a solar–mass TTS rotating at 20 km s^{-1}, the corotation radius at which the material in the Keplerian disk has the same angular velocity than the stellar surface lies at 6 $R_\star$. This is comparable to the value of the Alfvén radius derived by Bertout, Basri, and Bouvier (1988) assuming a plausible magnetic field strength of a few times 10^3 G at the stellar surface. Hence, magnetic fields may play an important role in controlling the angular momentum flow near the stellar surface during the accretion process.

IV. Rotational evolution to the Main Sequence

According to current models of pre–main sequence evolution, TTS will reach the main sequence as dwarfs with spectral types between A and M. Comparison between the rotation rates of TTS and of A–M dwarfs is thus expected to bring insight into the evolution of stellar rotation prior to the main sequence. Unfortunately, there are very few known low–mass stars with an intermediate evolutionary status between the T Tauri phase and the main sequence residence ("post–T Tauri star "). As a result, the pre–main sequence rotational history of low–mass stars cannot be traced observationally and theoretical models were developed in order to fill the observational gap between

TTS and dwarfs rotation. These models are constrained by two boundary conditions: on one extreme, the *vsini* distribution of TTS, which defines the initial conditions of pre–main sequence rotational evolution, and, on the other, the *vsini* distribution of dwarfs recently arrived onto the main sequence. Current observational and theoretical problems related to pre–main sequence rotational evolution are reviewed below.

a. High–mass stars ($1.25 \leq M_\star \leq 2.5\ M_\odot$)

TTS in this mass range will end onto the main sequence as A and early F dwarfs. Because they lack deep surface convective zones where stellar magnetic fields are amplified by a dynamo process, A and early F dwarfs are not expected to experience significant angular momentum loss during their evolution onto the main sequence. Hence, most field stars in this mass range are thought to have rotation rates much similar to those they had at the start of their main sequence evolution.

The mean rotation rate of dwarfs decreases from 150 km s^{-1} at spectral type A0 to 40 km s^{-1} at spectral type F3 (Fukuda 1982). TTS in the same mass range have *vsini* between 15 and 80 km s^{-1}, with a mean rotation rate of approximately 30 km s^{-1}. Assuming that *vsini* varies as $1/R_*$ during the star's contraction to the main sequence, i.e., that there is no angular momentum transfer across radial shells in these predominantly radiative stars, TTS will reach the ZAMS with *vsini* between 40 and 150 km s^{-1}, in good agreement with the mean rotation rates observed for A and F dwarfs (Bouvier *et al.* 1986, Hartmann *et al.* 1986). If, instead, solid–body rotation is assumed, TTS will end onto the main sequence with rotation rates much higher than those observed for dwarfs unless strong angular momentum loss occurs (Hartmann *et al.* 1986).

Although, on the average, A dwarfs have large rotational velocities, their *vsini* distribution is quite broad and exhibits an excess of slow rotators having *vsini* $\leq$ 40 km s^{-1}. If the above picture is correct, the progenitors of slowly–rotating A dwarfs would be pre–main sequence stars with rotation rates less than about 15 km s^{-1}, i.e., much lower than those commonly measured for TTS in the relevant mass range. There is growing suspicion, however, that most, if not all, slowly–rotating A stars belong to peculiar subgroups of the class, such as Ap and Am stars, or are members of short–period spectroscopic binaries, and have experience strong rotational braking (e.g, Fukuda 1982, Ramella *et al.* 1989). The low rotation rates of Ap stars is thought to result from magnetic braking prior and on the main sequence (Wolff 1981), while tidal interactions in close binaries may efficiently convert rotational into orbital motion (Wolff, Edwards, and Preston 1982). Although the mechanisms responsible for the low rotation rates of A stars with peculiar chemical abundances are still being debated, a large fraction of Am stars appear to be spectroscopic binaries in which tidal braking has occurred (Abt 1961). Conceivably, magnetic and tidal braking may prevent at least some TTS from spinning up during their contraction toward the main sequence. Whether these two mechanisms alone can account for the excess of slow rotators among A stars is however unknown.

b. Low–mass stars ($M_\star < 1.25\ M_\odot$)

It takes approximately 30 million years for a solar–mass TTS to reach the main sequence. By that time, it has usually moved far away from its parental cloud and is difficult to locate among more evolved dwarfs of the field. In some cases, however, stars have small enough space motions to remain close to each other over much longer time scales. Then, as the parental gas cloud rapidly dissipates, an open cluster remains which contains large amounts of roughly coeval stars whose distances and ages are well determined. Indeed, young open clusters offer the opportunity to study the rotational properties of large samples of stars recently arrived onto the main sequence.

Until the early 80's, the opinion prevailed that the rotational velocities of low–mass stars steadily decreased with time as a result of rotational braking due to magnetically-channeled mass–loss. This belief received support from the slow decline of rotation with age observed in cool main sequence stars (Kraft 1967, Skumanich 1972) and was still consistent with the more recent finding that low–mass TTS rotate much faster, on the average, than red dwarfs, which usually have $v\sin i \leq 5$ km s^{-1} by the age of the Sun. It thus came as a surprise when van Leeuwen and Alphenaar (1982) discovered K dwarfs in the young Pleiades cluster with rotational periods of a few hours, i.e., rotational velocities in excess of 150 km s^{-1}, which implied a much more complex rotational evolution than previously envisioned.

Subsequent studies of stellar rotation in young open clusters showed that rapid rotation was a widespread phenomenon among low–mass stars starting their evolution onto the main sequence (Stauffer *et al.* 1984, Stauffer, Hartmann, and Burnham 1985, Stauffer and Hartmann 1987, Stauffer *et al.* 1989). This is illustrated in Figure 4 (from Stauffer 1987), where the rotational velocities of low–mass stars in 3 open clusters are plotted against spectral type. From top to bottom, each panel displays stars belonging to open clusters of increasing age, thus providing a time sequence of the evolution of rotation in low–mass stars.

At an age of 50 million years, α Persei is the youngest open cluster for which significant amount of rotational data has been obtained (Stauffer, Hartmann, and Burnham 1985). In this cluster, G stars are already on the main sequence while lower mass stars are still slightly above it. The $v\sin i$ distribution of low–mass stars in the α Persei cluster is very wide, with approximately half of them rotating at less than 10 km s^{-1} while the others are more rapid rotators with $v\sin i$ up to more than 150 km s^{-1}.

If TTS with a mass less than 1.25 $M_\odot$ are the progenitors of late–type dwarfs in the α Persei cluster, the existence of rapid rotators in this cluster implies that at least some TTS must be strongly accelerated as they contract to the main sequence. Spin–up along radiative pre–main sequence tracks is expected to result both from the star's contraction and from the development of a radiative core in the stellar interior. Assuming solid-body rotation and no angular momentum loss, Stauffer *et al.* (1984) estimated that a 0.75 $M_\odot$ TTS will reach the main sequence with a 3–fold increase in rotational velocity. Thus, low–mass TTS will arrive on the main sequence with rotational velocities in the range of about 30 to 140 km s^{-1}, in rough agreement with the rotation rates observed for rapid rotators in the α Persei and Pleiades clusters (Hartmann *et al.* 1986, Bouvier *et al.* 1986).

Another explanation for the existence of rapid rotators among late–type dwarfs in young clusters would be that accretion from a circumstellar disk continues well beyond

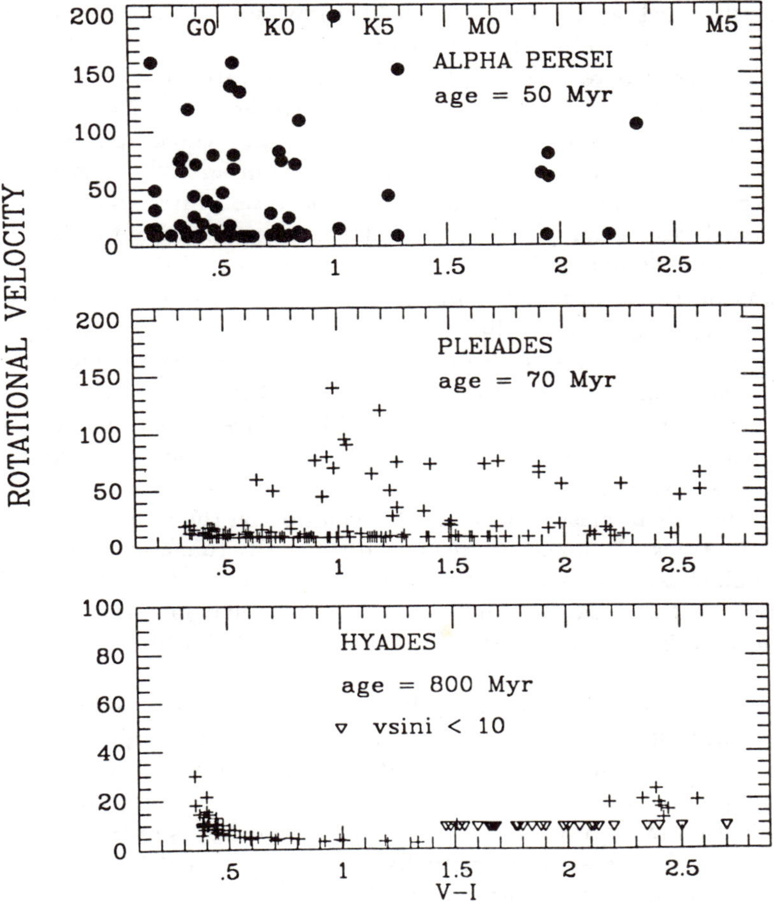

Figure 4. Rotational velocities of stars less massive than 1.3 $M_\odot$ in 3 young open clusters (from Stauffer 1987). The cluster ages are indicated in each panel, and the correspondence between (V–I) photometric index and spectral type is given in the upper panel. For the Alpha Persei and Pleiades clusters, spectroscopic velocities are shown and stars plotted at $v\sin i$ = 10 km s^{-1} only have upper limits set on their spectroscopic velocity. For most stars with (V–I) less than 1.4 in the Hyades cluster, true equatorial velocities were derived from rotational modulation studies, while only spectroscopic velocities were measured for less massive stars. Upper limits of 10 km s^{-1} on the spectroscopic velocity are shown as inverted triangles. Note the change of vertical scale in the lower panel.

the T Tauri stage, as low–mass stars approach the main sequence on radiative tracks. Then, despite angular momentum loss, stars which accrete from a disk would be spun up and arrive on the main sequence as rapid rotators, while stars with no accretion disks will be magnetically braked and arrive onto the main sequence as slow rotators. CTTS would thus be the progenitors of rapid rotators and WTTS those of slow rotators. This mechanism would explain both the high rotation rates observed in young cluster stars and the coexistence of slow and rapid rotators at any mass among these stars. However, the assumption that disk accretion continues all the way down to the main sequence is not supported by current observational data. If low–mass pre–main sequence stars on their radiative tracks were accreting from a disk at a rate close to 10^{-7} $M_{\odot}$ yr^{-1}, they would be expected to exhibit properties similar to those of CTTS and, therefore, would probably not have escaped detection. No such star, located approximately midway between the T Tauri phase and the main sequence in an H–R diagram, has been discovered so far (see Figure 1). Obviously, lower accretion rates ($\leq 10^{-8}$ $M_{\odot}$ yr^{-1}) would lead to much less striking properties but the existence of optically thick disks extending down to the stellar surface would still be betrayed at IR wavelengths by the thermal emission of warm dust in the inner disk regions. Even in the total absence of accretion, reprocessing of the stellar photons by dust in the disk would lead to a significant IR excess. Strom *et al.* (1989b) and Skrutskie *et al.* (1990) find that approximately half of pre–main sequence stars younger than 3.10^{6} years exhibit continuous energy excesses at 2.2 and 10 μm, while less than 10% of pre–main sequence stars older than 10^{7} years show significant near–IR excesses. They conclude that the lifetime of optically thick disks extending down to the stellar surface does not exceed 10^{7} years. Hence, by the time a solar–mass star begins its evolution on the radiative portion of its pre–main sequence track, the inner disk regions are optically thin and essentially clear of material. Sub–millimeter observations of pre–main sequence stars provide similar clues for disk evolution, in the sense that most of the circumstellar material surrouding the oldest stars is located in the outer parts of the disk (r $\geq$ 1 AU, Chini 1989, Beckwith *et al.* 1990).

Since accretion from a disk seems to be ruled out as a possible mechanism for spinning up low–mass stars on their radiative tracks, the observed spin–up must result from angular momentum conservation as the star contracts toward the main sequence. This is somewhat surprising because TTS winds are expected to carry significant amount of angular momentum away from the star. Hartmann and Noyes (1987) proposed that rotational braking in these stars could actually be quite inefficient if the magnetic field structure was chaotic rather than organized. Some support to this hypothesis comes from the comparison between the results of spectropolarimetric observations, which fail to detect magnetic fields at a level of a few hundred gauss in TTS (Brown and Landstreet 1981, Johnstone and Penston 1986, 1987), and the study of TTS magnetic activity, which provides indirect evidence for magnetic field strength of a few thousand gauss at the surface of TTS (Bouvier and Bertout 1989, Bouvier 1990). These two results can be reconciled by assuming that strong magnetic fields of opposite polarities exist at the stellar surface. Such a chaotic magnetic structure would remain unnoticed by polarimetric measurements which are only sensitive to the longitudinal field strength integrated over the stellar disk and would thus yield a null averaged value.

While angular momentum conservation on radiative tracks may account for α Persei rapid rotators, an approximately equal number of stars in this cluster have $v\sin i$ less than 10 km s^{-1}. Three broad classes of mechanisms have been proposed to date to account for these slow rotators: i) a large spread in angular momentum among pre–main sequence stars, ii) a spread in age among stars in young clusters, and iii) non–conventional pre–main sequence rotational braking.

Hartmann and Stauffer (1989) suggested that the progenitors of slowly–rotating α Persei stars should be searched for among TTS for which only $v\sin i$ upper limits of the order of 10 km s^{-1} could be derived. Indeed, these stars must rotate at a few km s^{-1} at most if they are to reach the main sequence with *vsini* less than 10 km s^{-1}. However, only 25% or so low–mass TTS have a *vsini* upper limit of 10 km s^{-1}. Moreover, Bouvier *et al.* (1986) measured *vsini* for 28 TTS with a resolution limit of a few km s^{-1} and failed to find any TTS with a *vsini* less than 6 km s^{-1}†. Therefore, it seems very unlikely that very slowly–rotating TTS exist in such a number as to explain the large fraction of slow rotators in young clusters. Alternatively, Bouvier *et al.* (1986) suggested that very slowly–rotating pre–main sequence stars may not exhibit T Tauri characteristics and therefore have escaped detection. Complete surveys of the stellar content of known formation regions, including those stars which show no emission, are needed to test this hypothesis.

A combination of rapid main sequence spin–down and significant age spread among late–type dwarfs in young clusters has been proposed by Stauffer *et al.* (1984) as an explanation for the coexistence of slow and rapid rotators in the same mass range. Comparison between the *vsini* distributions of stars in the α Persei and in the Pleiades clusters in Fig. 4 shows that all rapidly–rotating G stars in the α Persei cluster have been braked down to rotational velocities less than 20 km s^{-1} by the age of the Pleiades, i.e., in less than 20 million years after their arrival onto the main sequence (Stauffer, Hartmann, and Burnham 1985). This observational result agrees with models of the rotational evolution of solar–mass stars (Endal and Sofia 1981, Pinsonneault *et al.* 1989) which predict that, at the start of main sequence evolution, magnetic braking primarily affects the outer convective envelope while the radiative core remains in rapid rotation. Since the convective envelope contains only a small fraction of the total moment of inertia of a solar–mass star, rapid spin–down of the surface layers results. Stauffer, Hartmann, and Burnham (1985) further suggested that the braking time scale increases in lower mass stars as the convective envelope thickens. This would explain why a large fraction of K stars in the Pleiades cluster are still rapid rotators while all G stars have already been spun down (see Fig. 4). In the much older Hyades cluster, only the least massive stars are left with significant rotational velocities.

Assuming an age spread among the α Persei cluster stars comparable to the spin–down time scales of G and K stars, slow rotators would be the older stars which have first reached the main sequence where they were rapidly spun down, while rapid rotators would be younger and have not had time yet to experience significant main sequence braking. In order to test this hypothesis, Butler *et al.* (1987) measured lithium abundances in 4 rapidly–rotating and 4 slowly–rotating Pleiades K stars. They found that rapid rotators have an order of magnitude higher lithium abundances than slow rotators and interpreted this result as an age effect, indicating that rapid rotators are younger than slow ones. Whether lithium abundance is a good indicator of age in these stars is, however, questionable. In fact, the origin of the large spread observed in

---

† The lowest *vsini* upper limit measured so far, *vsini* $\leq$ 3 km s^{-1}, was reported by Bouvier (1990) for each component of the weak emission–line spectroscopic binary V826 Tau. While this measurement confirms previous suspicion that the binary system is seen almost pole–on (Mundt *et al.* 1983), it also provides an upper limit for the contribution of broadening agents other than rotation (e.g., macroturbulence, magnetic fields) to the width of the line profile.

the lithium abundances of pre–main sequence and of young cluster stars is poorly understood (Basri, Bertout, and Martin 1990, Balachadran, Lambert, and Stauffer 1988, Stauffer *et al.* 1989). In particular, clear understanding of the impact stellar rotation may have upon the abundances of light elements in low–mass stars is needed before any firm conclusion can be drawn from Butler *et al.*'s results.

The age spread hypothesis actually faces several difficulties. First, in order to account for the existence of slow rotators among α Persei K stars, an age spread of the order of the spin–down time scale of K stars is required, i.e., several 10^7 years, which is comparable to the cluster's age. The situation is even worse for slowly–rotating M stars, since the spin–down time scale for these stars seems to be of at least 10^8 years. Second, the recent study of a cluster even younger than α Persei, IC 2391 with an age of 30 million years (Stauffer *et al.* 1989), shows that low–mass stars in this cluster have $v\sin i$ in the range from 15 to 150 km s^{-1}, i.e., not significantly different from the $v\sin i$ distribution observed for low–mass stars in the Pleiades cluster. Yet, low–mass members of IC 2391 exhibit small dispersion about the 3.10^7 year isochrone, which indicates an age spread of 2.10^7 years at most. This result led Stauffer *et al.* (1989) to question the validity of the age spread hypothesis as an explanation for the wide $v\sin i$ distribution observed for low–mass stars in young clusters.

The third class of models proposed to account for slow rotators in young clusters refers to non–conventional rotational braking. Usual braking laws predict that the angular momentum loss rate scales with some power of the stellar rotational velocity. This is because angular momentum loss due to magnetically–channeled mass–loss is partly controlled by the magnetic field strength which, in turn, is assumed to scale with the star's rotational velocity. Then, rapid rotators are more strongly braked than slow ones and this results in a narrowing of the initial rotational distribution. Such braking laws are, therefore, unable to account for the observed widening of the rotational velocity distribution of low–mass stars between the T Tauri phase and the main sequence. As an alternative, Stauffer and Hartmann (1987) showed that the rotational velocity distribution of low–mass TTS could be reconciled with that of K and M Pleiades dwarfs assuming that rotational braking is independent of rotation, at least for stars with $v\sin i$ larger than 10 km s^{-1}. They justify this *ad hoc* assumption by noting that manifestations of magnetic activity in rapidly–rotating late–type stars, such as chromospheric and coronal emissions, weakly scale with rotation. This result, however, is somewhat controversial as Bouvier (1990) found evidence for a correlation between X–ray emission and angular rotation in TTS, which is similar to that found for late–type dwarfs and evolved binaries. More important, a key assumption of the models which account for initial fast rotation on the main sequence and subsequent rapid spin–down (e.g. Endal and Sofia 1981), is that angular momentum loss rate strongly depends upon rotation. Whether the alternative braking law proposed by Stauffer and Hartmann (1987) would still account for both pre–main sequence spin–up and rapid main sequence spin–down as observed is unclear.

None of the proposals summarized above consider the potentially important effects of binarity and/or planetary formation upon the early rotational evolution of low–mass stars. Recent observational studies of radial velocity variations in TTS suggest that the frequency of short–period binaries ($P < 100$ days) is similar among WTTS and solar–type dwarfs, and perhaps somewhat lower in CTTS (e.g., Mathieu, Walter, and

Myers 1989). Obviously, the pre–main sequence rotational history of single and binary stars may be quite different. Tidal torques in close binaries will maintain synchronism between rotational and orbital motions, so that angular momentum loss in these systems will result in a shrinkage of the orbits and subsequent spin–up of the binary components (see, e.g., van't Veer and Maceroni 1989). One may thus suspect that rapidly–rotating stars in young clusters are members of close binary systems while slow rotators are single stars. However, Stauffer *et al.* (1984) find no evidence for the radial velocity variations that would be expected if rapidly–rotating Pleiades K stars were short–period binaries. Moreover, it would be difficult to understand how rapidly–rotating close binaries manage to spin down to rotational velocities of 10 km s^{-1} or less shortly after their arrival onto the main sequence.

Alternatively, if no angular momentum loss occurs during pre–main sequence evolution, as seems to be required to account for fast rotators, members of close binaries will effectively be spun down. The reason is that, as stars in binary systems tend to spin–up due to the reduction of their moment of inertia on radiative tracks, tidal interactions will convert angular momentum of spin into orbital motions, which will result in wider orbits and further rotational braking until tidal torques are not strong enough to ensure synchronization between rotational and orbital periods. Indeed, tidal braking as been invoked by various authors to account for the excess of slow rotators among A–type dwarfs (see above). This mechanism would then identify rapid rotators in young clusters with single stars and slow ones with binary systems. This is consistent with Stauffer and Hartmann's (1987) finding that slow rotators among Pleiades K dwarfs are more often photometric binaries than rapid ones. While attractive, this hypothesis should be tested quantitatively by working out the time scales associated with angular momentum transfer in close binaries.

Transfer of angular momentum of spin into orbital motions may also take place during planetary formation. Stauffer and Hartmann (1986) and Stauffer and Soderblom (1989) speculated that slow rotators in young clusters could be stars which deposited most of their angular momentum into planetary systems while rapid rotators would have conserved their initial angular momentum. Current theoretical understanding of the dynamical evolution of proto–planetary disks around young stars does not allow one yet to test this hypothesis on quantitative grounds (see, e.g., Morfill and Wood 1989 and references therein). Apart from the Sun, there is only one clear–cut case of a main sequence star surrounded by a circumstellar disk. Direct CCD imaging of the star β Pictoris, a $2M_\odot$ dwarf whose age is estimated to be of a few times 10^6 years (R. Cayrel, priv. comm.), has revealed the existence of an optically thin dusty disk viewed nearly edge–on and extending more than 1100 AU away from the star (Smith and Terrile 1987). Whether large solid bodies have already formed in the disk is unknown. Nevertheless, the rotational velocity of β Pictoris, $v \simeq v\sin i = 139$ km s^{-1}, is quite typical of an A5 dwarf and actually corresponds to the peak of the rotational velocity distribution for stars in this mass range. This suggests that the rotational evolution of β Pictoris has not been significantly affected by the presence of a circumstellar disk and, perhaps, planetary formation.

V. Conclusion

In the last 10 years, the study of rotation in pre–main sequence stars has provided a number of severe observational constraints to be faced by theoretical models of star formation and early stellar evolution. Well-established observational facts may be summarized as follow: i) pre–main sequence stars rotate at a fraction of break-up velocity; ii) the increase of mean rotational velocity with mass observed in main sequence stars is already present, though less pronounced, in pre–main sequence stars; iii) weak and strong emission–line stars have similar rotational velocity distributions; iv) the distribution of rotational velocities in low–mass stars widens between the T Tauri phase and the arrival upon the main sequence, with at least some young stars being accelerated on their radiative pre–main sequence tracks. Various attempts have been made to address these results. Although models of star formation are not yet able to describe the complete evolutionary process from diffuse interstellar clouds to visible pre–main sequence stars, it clearly appears that rotational braking by magnetic fields must play a crucial role in the evolution of angular momentum during this phase. Parametrized models of the evolution of angular momentum in solar–mass stars from the T Tauri phase to the main sequence have reached qualitative agreement with observations in predicting both pre–main sequence spin–up and rapid main sequence spin–down. However, no satisfactory explanation has been proposed so far of the observed widening of the rotational velocity distribution of low–mass stars between the T Tauri phase and the zero–age main sequence.

While significant advances in our understanding of the evolution of angular momentum prior to the main sequence are to be expected from theoretical models in the present decade, several issues have still to be addressed observationally. In particular, observational efforts should be directed toward the determination of the rotation rates of TTS which only have upper limits of 10 km s^{-1} set on their spectroscopic velocity. Long–term photometric monitoring of these stars would allow one to measure rotational velocities down to a few km s^{-1}, thus providing a complete determination of TTS rotational velocity distribution. Another critical issue relates to the binary frequency among TTS. On–going efforts made by several groups using various observational techniques (CCD imaging, optical and IR speckle interferometry, lunar occultations, radial velocity variations) should soon provide reliable estimates of the fraction of binaries among pre–main sequence stars. Finally, the very high signal–to–noise ratio required to measure the Zeeman splitting of photospheric line profiles at high spectral resolution might already be obtainable for the brightest TTS. Obviously, since magnetic fields are thought to play a crucial role in the rotational evolution of young stars, a direct determination of this key parameter is highly desirable.

Acknowledgements: Discussions with M. Gerbaldi and J. Zorec on the rotational properties of A and B dwarfs are gratefully acknowledged. I am also indebted to J. Stauffer for providing Figure 4 of this paper.

References

Abt, H.A 1961, *Astrophys. J. Supp. Series*, **52**, 37
Balachadran, S., Lambert, D.L., and Stauffer, J.R. 1988, *Astrophys. J.*, **333**, 267
Basri, G., and Bertout, C. 1989, *Astrophys. J.*, **341**, 340
Basri, G., Bertout, C., and Martin E. 1990, *preprint*
Beckwith, S.V.W., Sargent, A.I., Chini, R.S., and Güsten, R. 1990, *preprint*
Bertout, C. 1989, *Ann. Rev. Astron. Astrophys.*, **27**, 351
Bertout, C., Basri, G., and Bouvier, J. 1988, *Astrophys. J.*, **330**, 350
Bertout, C., Basri, G., and Cabrit, S. 1990, in *The Sun in Time*, eds M.S. Giampapa and G.R. Sonnet (Univ. of Arizona Press, Tucson), in press
Bodenheimer, P. 1978, *Astrophys. J.*, **224**, 488
Boss, A.P. 1986, *Astrophys. J. Supp. Series*, **62**, 519
Bouvier, J. 1990, *Astron. J.*, **99**, 946
Bouvier, J., and Bertout, C. 1989, *Astron. Astrophys.*, **211**, 99
Bouvier, J., Bertout, C., Benz, W., and Mayor, M. 1986, *Astron. Astrophys.*, **165**, 110
Brosche, P. 1986, *Comments in Astrophys.*, Vol.11, No.5, p.213
Brown, D.N., and Landstreet, T.D. 1981, *Astrophys. J.*, **846**, 299
Butler, R.P., Cohen, R.D., Duncan, D.K., and Marcy, G.W. 1987, *Astrophys. J. Letters*, **319**, L19
Cabrit, S., Edwards, S., Strom, S.E., and Strom, K.E. 1990, *Astrophys. J.*, in press
Chini, R. 1989, in *Low-Mass Star Formation and Pre-Main Sequence Objects*, ed. B. Reipurth (ESO: Garching), p.173
Cohen, M., and Kuhi, L.V. 1979, *Astrophys. J. Supp. Series*, **41**, 743
Draine, B.T. 1983, *Astrophys. J.*, **270**, 519
Durisen, R.H., Gingold, R.A., Tohline, J.E., and Boss, A.P. 1986, *Astrophys. J.*, **305**, 281
Endal, A., and Sofia, S. 1981, *Astrophys. J.*, **243**, 625
Finkenzeller, U. 1984, *Astron. Astrophys.*, **151**, 340
Fukuda, I. 1982, *Publ. Astron. Soc. Pac.*, **94**, 271
Ghosh, P., and Lamb, F.K. 1979, *Astrophys. J.*, **234**, 296
Goldsmith, P.F., and Arquilla R. 1985, in *Protostar and Planets*, eds. D.C. Black and M.S. Matthews, Vol.II, p.137
Hartmann, L.W., and MacGregor, K.B. 1982, *Astrophys. J.*, **259**, 180
Hartmann, L.W., and Noyes, R.W. 1987, *Ann. Rev. Astron. Astrophys.*, **25**, 271
Hartmann, L.W., and Stauffer, J.R. 1989, *Astron. J.*, **97**, 873
Hartmann, L.W., Soderblom, D.R., and Stauffer, J.R. 1987, *Astron. J.*, **93**, 907
Hartmann, L., Hewett, R., Stahler, S., and Mathieu, R.D, 1986, *Astrophys. J.*, **309**, 275
Herbig, G.H. 1960, *Astrophys. J. Supp. Series*, **4**, 337
Herbig, G.H., and Bell K.R. 1988, *Lick Observatory Bulletin No. 1111*
Herbig, G.H., Vrba, F.J., and Rydgren, A.E. 1986, *Astron. J.*, **91**, 575
Heyer, M.H. 1988, *Astrophys. J.*, **324**, 311
Heyer, M.H., Vrba, F.J., Snell, R.L., Schloerb, F.P., Strom, S.E., Goldsmith, P.F., and Strom, K.M. 1987, *Astrophys. J.*, **321**, 855
Johnstone, R.M., and Penston, M.V. 1986, *Monthly Not. Roy. Astron. Soc.*, **219**, 927
Johnstone, R.M., and Penston, M.V. 1987, *Monthly Not. Roy. Astron. Soc.*, **227**, 797
Joy, A.H. 1942, *Publ. Astron. Soc. Pac.*, **54**, 15
Joy, A.H. 1945, *Astrophys. J.*, **102**, 168
Kraft, R.P. 1967, *Astrophys. J.*, **150**, 551
Kraft, R.P. 1970, in *Spectroscopic Astrophysics*, ed. G.H. Herbig (Univ. of California Press), p. 385

Lada, C.J. 1985, *Ann. Rev. Astron. Astrophys.*, **23**, 367
Mathieu, R.D., Walter, F.M., and Myers, P.C. 1989, *Winsconsin Astrophys. Preprint*, No.320
Mestel, L., and Spitzer, L. 1956, *Monthly Not. Roy. Astron. Soc.*, **116**, 503
Morfill, G.E., and Wood, J.A. 1989, *Icarus*, **82**, 225
Mouschovias, T.Ch. 1977, *Astrophys. J.*, **211**, 147
Mouschovias, T.Ch., and Paleolougo, E.V. 1979, *Astrophys. J.*, **230**, 204
Mouschovias, T.Ch., and Paleolougo, E.V. 1980, *Astrophys. J.*, **237**, 877
Mouschovias, T.Ch., and Paleolougo, E.V. 1986, *Astrophys. J.*, **308**, 781
Mundt, R., Walter, F.M., Feigelson, E.D., Finkenzeller, U., Herbig, G.H., and Odell, A.P. 1983, *Astrophys. J.*, **269**, 229
Nakano, T. 1979, *Publ. Astron. Soc. Japan*, **31**, 697
Norman, C., and Heyvaerts, J. 1985, *Astron. Astrophys.*, **147**, 247
Pinsonneault, M.H., Kawaler, S.D., Sofia, S., and Demarque, P. 1989, *Astrophys. J.*, **338**, 424
Pudritz, R.E., and Norman, C.A. 1986, *Astrophys. J.*, **301**, 571
Ramella, M., Gerbaldi, M., Faraggiana, R., and Böhm, C. 1989, *Astron. Astrophys.*, **209**, 233
Schatzman, E. 1962, *Annales d'Astrophysique*, **25**, 18
Shu, F.H., Adams, F.C., and Lizano, S. 1987, *Ann. Rev. Astron. Astrophys.*, **25**, 23
Skumanich, A. 1972, *Astrophys. J.*, **171**, 565
Skrutskie, M.F., Dutkevitch, D., Strom, S.E., Edwards, S., Strom, K.M., and Shure, M.A. 1990, *Five Coll. Astron. Preprint Ser.*, No.721
Smith, B.A., and Terrile, R.J. 1987, *B.A.A.S.*, Vol.19, No.3, 829
Spitzer, L. 1978, *Physical Processes in the Interstellar Medium*, J. Wiley & Sons Eds.
Stauffer, J.R. 1987, *Lect. Notes in Phys.*, **291**, 182
Stauffer, J.R., and Hartmann, L.W. 1986, *Publ. Astron. Soc. Pac.*, **98**, 1233
Stauffer, J.R., and Hartmann, L.W. 1987, *Astrophys. J.*, **318**, 337
Stauffer, J.R., and Soderblom, D.R. 1989, Proc. of *The Sun in Time* Conf., Tucson, Arizona, 1989, in press
Stauffer, J.R., Hartmann, L.W., and Burnham, J.N. 1985, *Astrophys. J.*, **289**, 247
Stauffer, J.R., Hartmann, L.W., Jones, B.F., and McNamara, B.R. 1989, *Astrophys. J.*, **342**, 285
Stauffer, J.R., Hartmann, L.W., Soderblom, D.R., and Burnham, N. 1984, *Astrophys. J.*, **280**, 202
Strom, K.M., Wilkin, F.P., Strom, S.E., and Seaman R.L. 1989a, *Astron. J.*, **98**, 1444
Strom, K.M., Strom, S.E., Edwards, S., Cabrit, S., and Strutskie, M.F. 1989b, *Astron. J.*, **97**, 1451
Terebey, S., Shu, F.H., and Cassen, P. 1984, *Astrophys. J.*, **286**, 529
Trimble, V. 1984, *Comments in Astrophys.*, Vol.10, No.4, p.127
van Leeuwen, F., and Alphenaar, P. 1982, *ESO Messenger*, **28**, 15
van't Veer, F., and Maceroni, C. 1989, *Inst. Astrophys. Paris Preprint*, No. 292
Vogel, S.N., and Kuhi, L.V. 1981, *Astrophys. J.*, **245**, 960
Walter, F.M. 1986, *Astrophys. J.*, **306**, 573
Walter, F.M., Brown, A., Mathieu, R.D., Myers, P.C., and Vrba, F. 1988, *Astron. J.*, **96**, 297
Weaver, C.B. 1987, *Astrophys. J. Letters*, **319**, L89
Wolff, S.C. 1981, *Astrophys. J.*, **244**, 221
Wolff, S.C., Edwards, S., and Preston, G.W. 1982, *Astrophys. J.*, **252**, 322
Zorec, J. 1986, Thèse de Doctorat d'Etat, Univ. Paris VII

THE INTERNAL ROTATION OF THE SUN

Werner Däppen
Space Science Department of ESA, ESTEC
2200 AG Noordwijk, The Netherlands

Abstract. An introduction to current techniques to infer the Sun's internal rotation from observed acoustic oscillation modes is given, and some representative results are shown.

1. Introduction

As soon as the solar five-minute oscillations were recognized as a superposition of a large number of normal modes, their precisely determined oscillation frequencies became an important tool to diagnose the physics of the Sun's interior. The internal rotation has been an obvious candidate for helioseismology, since it leaves a qualitative imprint in the form of frequency splittings. In the absence of symmetry-breaking forces, the frequency of an oscillation mode of angular degree l is $(2l + 1)$-fold degenerate. Rotation lifts this degeneracy, and in the power spectrum multiplets with $2l + 1$ terms appear. The rotational splittings are tiny: the frequency separation is determined by the period of rotation, *i.e.* about 25 days, 7500 times larger than a typical p-mode period (5 minutes). Nevertheless, the spectacular observational progress of helioseismology of the last 15 years has made the observation of these rotational splittings possible, and an already quite clear picture of the internal rotation of the Sun is emerging (for recent reviews see, *e.g.*, Harvey, 1988; Christensen–Dalsgaard, 1989).

I will begin with a brief introduction to the theoretical principles (section 2) and observational facts (section 3) of helioseismology. Then, in section 4, I will discuss the connection between solar rotation and oscillation frequencies, both in the forward and backward direction.

The purpose of this article is to show the basic physical principles behind the recent developments. With this emphasis on methods, I will not be able to mention many details (for more see *e.g.* Christensen–Dalsgaard, 1989). However, I will explain in some detail those definitions that are used in the presentations of the observers (such as the Legendre expansion coefficients a_i's).

2. Helioseismology: theory

In this section, I present two different approaches to the dynamical problem of finding the solutions for the motion of the stellar fluid around its equilibrium state. Nonlinearities are neglected, and the problem of stellar oscillations is equivalent to finding all normal modes. The first approach will be an outline of the complete numerical solution. The similarity with other eigenvalue problems in mathematical physics is stressed (*e.g.* vibrating strings, bound states in quantum mechanics). The second approach will be based on a simplified wave equation and on propagation diagrams that reflect local conditions in a star. This approach is best suited for a qualitative discussion of the frequency spectrum of modes. Important characteristics of the modes, like their type (p or g mode, see below) or penetration depth are directly visible in propagation diagrams. For both approaches, my emphasis will be on methods. A much more detailed description can be found, for instance, in the recent review by Christensen–Dalsgaard and Berthomieu (1990).

2.1 *Outline of numerical computations of stellar oscillations*

Assuming the existence of a stable and constant equilibrium configuration, we start out from the hydrodynamic equations for compressible fluids

$$\frac{\partial \mathbf{v}}{\partial t} + \mathbf{v} \cdot \nabla \mathbf{v} = -\frac{1}{\rho} \nabla p + \nabla \psi \ . \tag{1}$$

Here, $\mathbf{v}$ is the (Eulerian) velocity field, p and ρ pressure and density, respectively, and ψ is the (self-) gravitational potential. For simplicity, we have disregarded viscosity. To this equation, one must add the usual equation of continuity and also an energy equation, which - in the simplest case - is replaced by a condition of adiabaticity, normally expressed in the form of constant co-moving specific entropy (per mass). Under the assumption of adiabaticity, stellar pulsation is frictionless and energy conserving. This condition is very well satisfied in the bulk of the solar interior. Only close to the surface does the thermal time scale become comparable to the dynamical one (of about one hour); deeper down the thermal relaxation time is much longer (about 10^7 years for the entire Sun).

It is clear that the assumption of adiabatic motion precludes any discussion of mode excitation and damping, necessary to understand why a given pulsation mode is active while another is not. Nevertheless, this hypothesis of adiabaticity still leads to many important results by telling, *e.g.*, what the linear eigenfrequencies are. Virtually the whole success of helioseismology has been so far in the framework of adiabatic pulsations; only very recently have serious attempts been made to go beyond and to address questions like mode excitation, damping, and amplitudes (for a review see Cox *et al.*, 1990).

In the following I discuss a few of the key steps used in manipulating Equation (1). I will be very brief, but details can be found, *e.g.*, in the book by Unno *et al.* (1989). In this section, we restrict ourselves to a simple, but still relevant case, in which the equilibrium configuration is assumed to be at rest and spherically symmetric (rotation will be discussed in chapter 4.). The gravitational potential is assumed to be static, *i.e.* its distortions due to the stellar pulsation itself is neglected (this is the so-called Cowling approximation, see Unno *et al.*, 1989). Restricting even more to *small* perturbations, we introduce *linearization* of Eq. (1).

The static equilibrium and the linearized equations allow the separation of the time dependence in the form of $\exp(i\omega t)$. The spherically symmetric equilibrium configuration allows expressing the field variables (displacement vector, pressure, and density) as a series of spherical harmonics Y_l^m of angular degree l and azimuthal order m, so that each term in the series is itself a solution of the equations. The adiabatic assumption is used to link pressure and density fluctuation with the help of a thermodynamical quantity (the adiabatic exponent), which is a given quantity of the equilibrium model. One arrives therefore at a fourth-order system for the (independent) three displacement-vector components and the pressure fluctuation. The Cowling approximation allows yet another simplification; with it, the tangential part of the displacement field becomes proportional to the tangential component of the gradient of the (Eulerian) pressure fluctuation, and thus the only independent fields remaining are the radial component of the displacement vector and the pressure fluctuation. Their amplitudes are governed by the following (schematic) system of ordinary differential equations

$$\begin{aligned} \frac{dy_1}{dr} &= f_{11}(r)y_1 + f_{12}(r;l;\omega^2)y_2 \\ \frac{dy_2}{dr} &= f_{21}(r;\omega^2)y_1 + f_{22}(r)y_2 \\ y_1 &\equiv \frac{\xi_r}{r} \\ y_2 &\equiv \frac{p'}{gr\rho} \ . \end{aligned} \tag{2}$$

Here, the functions $f_{ij}(r;\omega^2)$ are expressions involving quantities of the equilibrium model (like pressure, density, local gravity, sound-speed, etc.). As is standard practice, the labels l and ω are dropped in the y_i's. The prime $'$ denotes the Eulerian (first-order) displacement from equilibrium.

Adding boundary conditions to equation (2) leads to an eigenvalue problem. The one at the *center* is the usual regularity condition due to the singular nature of Eq. (2). From the specific nature of the coefficients one knows (see Unno *et al.*, 1989) that Eq. (2) has a regular-singular point at $r = 0$, and so there is a regular and a singular solution. Picking the regular one gives the boundary condition (this is explicitly done by a standard power-series development). The *outer* boundary condition is not so simple. In principle, one would have to put a good stellar atmosphere (for which there are elaborate models) at the outer end, and impose smooth matching as the outer boundary condition. Until now, nobody has done this in a satisfactory way, and simpler approaches must be chosen. Often an isothermal atmosphere is assumed, and the outer boundary condition is determined by a discussion of propagation and reflection of sound waves in a stratified atmosphere analogous to Lamb (1932) (we come to that in 2.2.). Here, we can afford something even simpler, namely a mechanical boundary condition of the form

$\delta p = 0$ where δ stands for the Lagrangian displacement. Such a boundary condition is the three-dimensional analogon of a frictionless lid. In terms of our variables y_1 and y_2, the condition $\delta p = 0$ is given by $y_1 + y_2 = 0$.

In principle, finding the eigenvalues of Eq. (2) is really not much harder than those of a vibrating string, whose equation is $d^2y/dx^2 + \omega^2 y = 0$ with boundary conditions $y = 0$ at two different x values. The only complication in Eq. (2) comes from the nonconstant coefficients, but numerically it is still an easy task. The hard part is finding the equilibrium solution which delivers the coefficients for Eq. (2). A still excellent introduction to the basic principles of stellar modelling is the book by Schwarzschild (1958) (though it cannot, of course, cover the fascinating progress that has been made since then, mainly thanks to the tremendous increase in computing power).

2.2 *Qualitative discussion using propagation properties*

We adopt the asymptotic discussion of Deubner and Gough (1984), which itself is similar to the treatment of acoustic waves by Lamb (1932) [see also Christensen-Dalsgaard (1986)]. For wavelengths much shorter than the solar radius, normal oscillation modes can be quite accurately discussed using the simplified wave equation

$$\Psi'' + K^2(r)\Psi = 0 \; . \tag{3}$$

Here, $\Psi = \sqrt{\rho}c^2 \mathrm{div}(\delta\mathbf{R})$, where ρ and c are density and sound speed of the equilibrium configuration, and $\delta\mathbf{R}$ is the fluid displacement vector. The local wave number is given by

$$K^2(r) = \frac{\omega^2 - \omega_c^2}{c^2} + \frac{l(l+1)}{r^2}\left(\frac{N^2}{\omega^2} - 1\right) , \tag{4}$$

with the acoustic cut-off frequency defined by

$$\omega_c^2 = \frac{c^2}{4H^2}(1 - 2\frac{dH}{dr}) , \tag{5}$$

and the Brunt-Väisälä frequency N by

$$N^2 = g(\frac{1}{H} - \frac{g}{c^2}) , \tag{6}$$

where H is the density scale height and g the local gravity. From the form of Eq. (3) (to which upper and lower boundary conditions must be added), one immediately realizes that in propagation zones necessarily $K^2 > 0$.
Our present qualitative discussion of the influence of mass and evolution on oscillation frequencies aims at showing the maximum of effects with a minimum of curves. Here, we restrict ourselves to the role of N^2 and ω_c^2 in K. For finer details we refer the reader to Deubner and Gough (1984), Christensen-Dalsgaard (1986) or Gough (1985).

With the convenient definition of the Lamb frequency

$$S_l^2 = \frac{l(l+1)}{r^2}c^2 , \tag{7}$$

we obtain the simplified necessary conditions for propagation of an acoustic wave, $\omega > \omega_c$ and $\omega > S_l$. Additionally, in order to have a trapped standing wave, it is also necessary that in some surface layer ω_c becomes greater than ω. This happens indeed; the height of this (outer) ω_c mountain is the greater, the cooler the local temperature at the edge of the star is [this is seen from Eq. (5)]. 'Mathematical' stars with zero temperature at the outer boundary have an infinitely high ω_c mountain; they can therefore trap modes of arbitrary high frequency. Real stars have an 'inversion' temperature (just above the photosphere); further up temperature begins to rise again. The maximum p-mode frequency 'measures' this inversion temperature.

In propagation zones (if a constant adiabatic exponent of 5/3 is assumed), a further simplification follows from the fact that $\omega > g/c$ implies $\omega > \omega_c$. And finally, we choose the approximation of identifying (the absolute value of) g/c with N, which certainly gives the correct order of magnitude in radiative zones (but would be entirely wrong in convection zones, where $N \approx 0$). The advantage of this choice is that the same simple curves will also give some information about g modes.

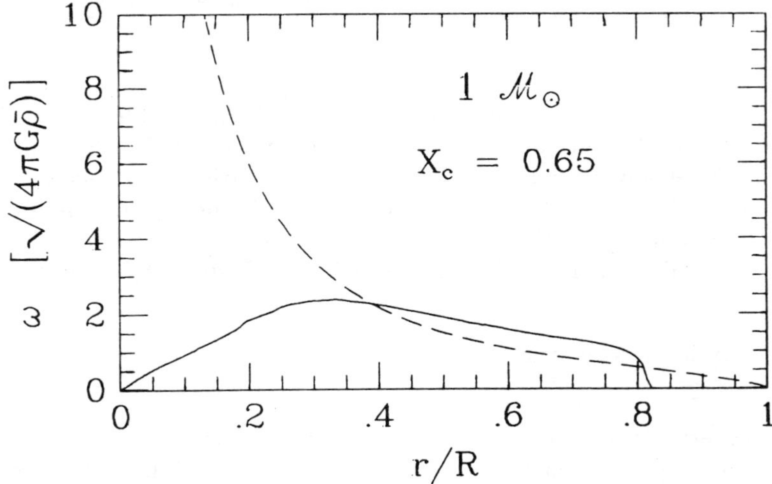

Figure 1. Critical frequencies as functions of the fractional radius r/R. The solid line denotes the Brunt-Väisälä frequency N, the dashed line the Lamb frequency S_l for $l = 1$. The model parameters are: hydrogen abundance $X = 0.70$, heavy-element abundance $Z = 0.01$, and the mixing-length parameter $l/H_p = 1.5$. Stellar age is indicated by the central hydrogen abundance X_c (from Däppen *et al.*, 1988).

Let us now consider S_l (we choose the representative case $l = 1$) and N in a model of a 1 $M_\odot$ star (Figure 1). Due to the rather deep convection zone, N cannot represent the increase of ω_c close to the surface, and so the diagram does not show the upper turning point that is caused by the large spatial inhomogeneity near the surface. In Figure 1, S_1 defines the penetration depth of the $l = 1$ modes; for $l > 1$ the corresponding curves would be shifted to the right, as required by Eq. (7).

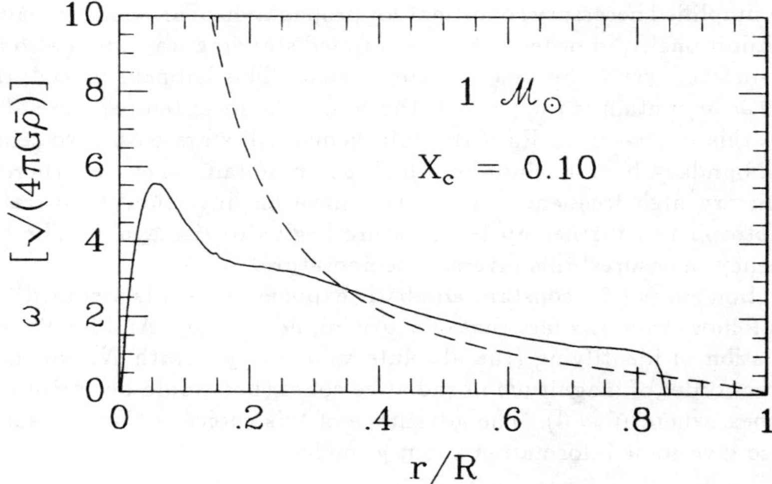

Figure 2. Same as Figure 1, but for a more evolved model (from Däppen *et al.*, 1988).

Figure 1 also shows how to distinguish p modes from g modes. The distinction is only asymptotic. Modes with $\omega \to \infty$ would become true p modes (but they cease to be trapped above a certain frequency, see above). Modes with $\omega \to 0$ become true g modes [they have a large K^2 due to N/ω, see Eq. (4)]. Outside the asymptotic regime one still speaks of p modes and g modes, but they are not 'pure', though most of them are of a dominant type. Only in highly evolved stars do genuine dual-status modes appear.

3. Helioseismology: observations

After the discovery of the solar 5-minute oscillations by Leighton *et al.* (1962), it took 15 years before they were recognized as global oscillations (Deubner, 1975; Rhodes *et al.*, 1977). To illustrate the tremendous observational progress made since, I oppose mid-seventies data (Figure 3) to end-eighties data (Figure 4).

Figure 3 shows one of the earliest so-called $k - \omega$ diagrams (contours of velocity power as a function of frequency ω and horizontal wavenumber k_h). Barely visible ridges connect modes with the same radial order n. These diagrams gave the final proof of the global solar nature of the 5-minute oscillations, because their qualitative features were those of any oscillating gaseous sphere, and because quantitatively they correspond to what we expect from the Sun (note the model predictions shown in Figure 3). The tremendous observational progress of helioseismology made since (intermediate steps were *e.g.* Duvall, 1982; Harvey and Duvall, 1984) is demonstrated by the state-of-the-art Figure 4 (from Libbrecht and Woodard, 1990). Note the 1000 σ error bars! Further note that the horizontal wave number k_h has been replaced by the angular degree l, which is more useful for the theorist.

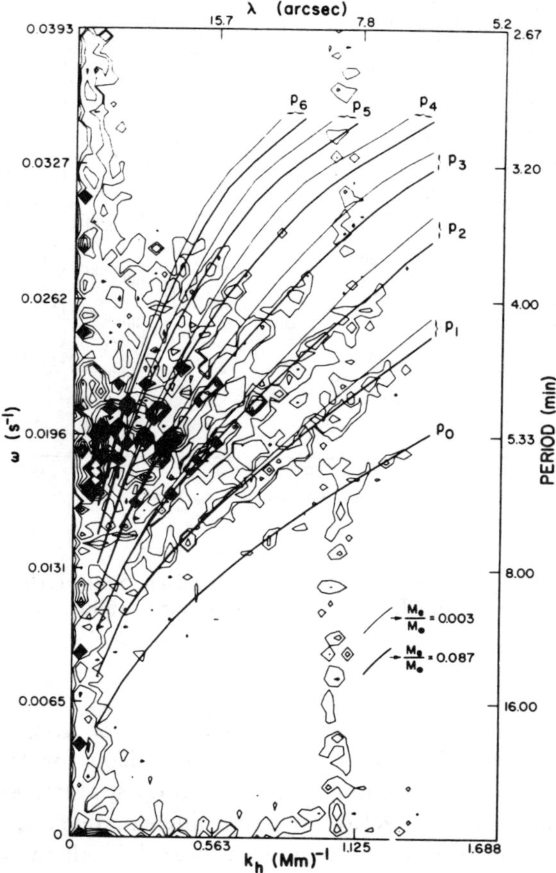

Figure 3. $(k_h - \omega)$-diagrams of velocity fluctuations. The contours indicate observed velocity power (in some units), and the solid lines are theoretical calculations for two models, with mass fractions of the convective envelope as indicated at the lower right (from Rhodes *et al.*, 1977).

The data like those of Figure 4 [see Libbrecht *et al.* (1990) for the most recent frequency tables] allow a high precision analysis of the structure of the solar interior. Indeed, the relative precision, with which each of the observed mode frequencies ν_{nl} is

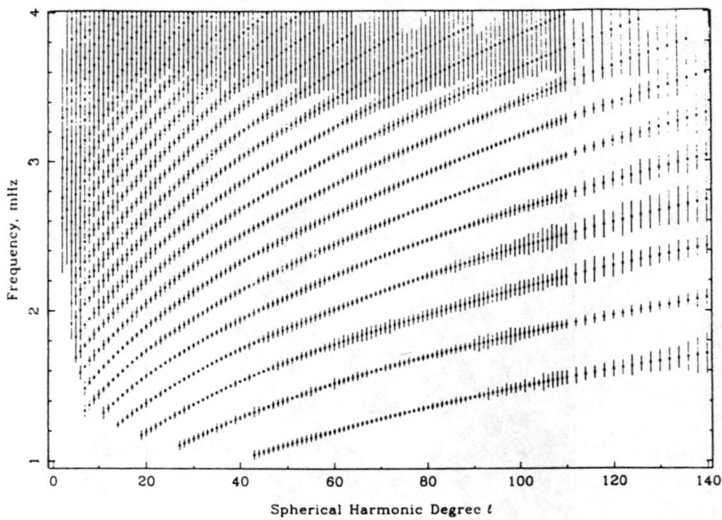

Figure 4. Observed 1986 p-mode frequencies from Big Bear Solar Observatory, with 1000 times magnified 1σ error bars (!) (from Libbrecht and Woodard, 1990).

determined, attains 10^{-4}, which is at least one order of magnitude better than the uncertainties of the theoretical predictions. The reason for this inadequacy of the theoretical models is that they are not (yet) sufficiently sophisticated, because the usual simplifying assumptions on convection, opacity, nuclear physics, internal rotation, and other physical ingredients are not good enough to explain all the details encountered in the seismological data.

If the Sun were spherically symmetric, then each mode frequency ν_{nl} would be $2l+1$ times degenerate. The solar rotation breaks this symmetry (like any other nonspherical perturbation, such as *e.g.* magnetic fields), thus it splits each frequency into a multiplet. This is a small effect, since the characteristic frequency splitting is of order of that of the solar rotation whose period is a little less than a month. Therefore the rotational splitting is too small to be visible in a plot of absolute frequencies such as Figure 4. However, thanks to observation series of weeks and months, the splittings can be well observed for a wide range of l (see, *e.g.*, Harvey, 1988 and references therein;

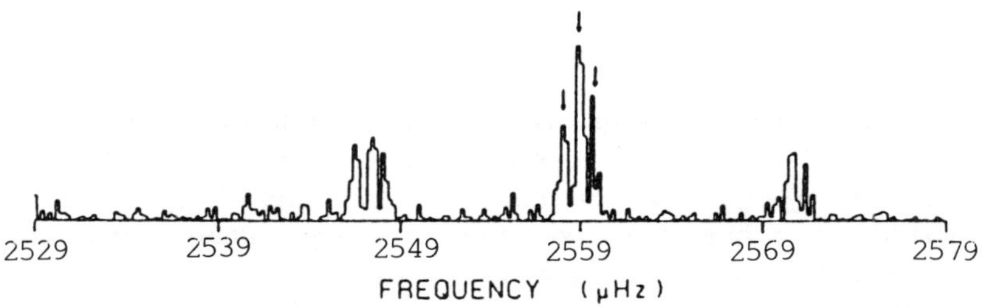

Figure 5. Rotational splitting of an $l = 1$ p mode ($n = 17$). The upper and lower side bands, separated from the real peak by 11.57 μHz (corresponding to the 24-hour modulation of the signal), are also visible (from Pallé *et al.*, 1988).

additional references are given in section 4). To illustrate, Figure 5 shows a typical rotational splitting.

Solar oscillations are measured either in intensity or velocity, with or without spatial resolution. Velocity measurements use the Doppler shift of a certain spectral line [see, for an introduction, the book by Unno *et al.*, (1989)]. Whole-disk observations (*e.g.* Claverie *et al.*, 1979) project the Sun onto a point; the absence of spatial resolution precludes observation of modes with l greater than about 3. In other observations the Sun is projected onto the equator line (Duvall and Harvey, 1984). Such observations can separate modes with different l, but not all m (see below). Only the extreme case $m = \pm l$ modes can so be detected. A complete identification of all modes, with any l and m, can be achieved with full-disk measurements; in these the solar disk is resolved into a large set of 'pixels', for each of which the relative velocity is measured (*e.g.* Brown, 1985; Duvall *et al.*, 1986; Brown and Morrow, 1987; Korzennik *et al.*, 1988; Rhodes *et al.*, 1988).

Before coming to the importance of the m dependence of mode frequencies, I should like to emphasize the smallness of the *amplitudes* of the solar oscillations. Individual modes have velocity fields on the order of $10 - 20$ cm/s, which correspond to amplitudes on the order of 10^{-7} relative to the solar radius. The only reason why the observational techniques (which are admittedly very sophisticated) can succeed in measuring these tiny oscillations is the strict periodicity of a mode. Typical mode lifetimes are from a few days to a couple of weeks. The fact that a mode can be followed over a very large number of periods allows to filter out solar and terrestrial noise (*e.g.* solar granulation, fluctuations of the earth's atmosphere, *etc.*).

As mentioned above, if the Sun were spherically symmetric, then each mode frequency ν_{nl} would be $2l + 1$ times degenerate. To first order, a rotationally split multiplet is equally spaced, and the deviation from this linear relation is small (see section 4). For this reason, the whole multiplet can adequately be described by a small set of parameters that define the spacing and the deviation from linearity,

$$\nu_{nlm} = c_{nl}^{(0)} + c_{nl}^{(1)} m + \frac{1}{2} c_{nl}^{(2)} m^2 + ... \tag{8}$$

The few lowest coefficients $c_{nl}^{(i)}$ then carry all information needed, allowing a substantial reduction in the volume of data, especially for the higher l. While in principle each multiplet ν_{nlm} $(m = -l, ... + l)$ determines a polynomial of degree $2l + 1$ and thus all the $2l+1$ coefficients c_i in a unique way, the purpose of expanding in a series of type (8) is to mimic the high-l multiplets to a good approximation by a much smaller number of parameters. In this case, however, there is no unique determination of the coefficients, and therefore there are better and worse expansions.

In practice, it has turned out that the following expansion in a set of *orthogonal* polynomials has been more adequate (Duvall *et al.*, 1986; see also the review by Harvey, 1988)

$$\nu_{nlm} - \bar{\nu}_{nl} = L \sum_{i=0}^{N} a_i P_i(-\frac{m}{L}), \tag{9}$$

where $\bar{\nu}_{nl}$ denotes the m-averaged frequency of the multiplet, P_i the Legendre polynomials of degree i, $L = \sqrt{l(l+1)}$, and, to minimize the variation of the coefficients with degree l, m/L rather than m/l is chosen as argument in P_i. Although, as said above, the full expansion (with $N = 2l + 1$) would be completely equivalent to equation (8), the truncated expressions reduce the dependence of the coefficients on each other and on the value of N. Current observational accuracy allows $N = 5$, see Harvey (1988). It has become customary to present and discuss rotational splittings in terms of these coefficients a_i (see following section).

4. Oscillation frequencies and internal rotation

4.1 *Forward problem*

4.1.1 Uniform rotation

In a rough approximation, the Sun rotates rather uniformly, and the sidereal equatorial rotation period is about 25 days. The observed differential rotation of the surface can be expressed by empirical laws (depending slightly on the method employed to measure the rotation; for an introduction, see the book by Tassoul, 1978). A typical, still fairly accurate expression for the surface differential rotation ζ (in degrees per day) as a function of heliocentric latitude ϕ is (Newton and Nunn, 1951; for more modern expressions, see *e.g.* Christensen–Dalsgaard, 1989)

$$\zeta = 14.38 - 2.77 \sin^2 \phi \quad [\text{deg/day}]. \tag{10}$$

Neglecting differential rotation for a moment, we transform Equation (1) into a uniformly rotating coordinate system

$$\frac{\partial \mathbf{v}}{\partial t} + \mathbf{v} \cdot \nabla \mathbf{v} + 2\boldsymbol{\Omega} \wedge \mathbf{v} + \boldsymbol{\Omega} \wedge (\boldsymbol{\Omega} \wedge \mathbf{r}) = -\frac{1}{\rho} \nabla p + \nabla \psi \ , \tag{11}$$

with $\boldsymbol{\Omega}$ denoting the (constant) angular velocity and $\wedge$ the (ordinary) vector cross product. Since $\boldsymbol{\Omega}$ is small, perturbation theory can be used to compute the frequency correction due to $\boldsymbol{\Omega}$ in terms of the unperturbed (complex) mode ξ_{nlm}

$$\begin{aligned} \delta\xi_{nlm}(r,\theta,\phi,t) \ &= \\ \left[\xi_r(r;n,l) Y_l^m \mathbf{e_r} + \xi_h(r;n,l) \left(\frac{\partial Y_l^m}{\partial \theta} \mathbf{e}_\theta \ + \ \frac{1}{\sin\theta} \frac{\partial Y_l^m}{\partial \phi} \mathbf{e}_\phi \right) \right] & e^{-i\omega_{nl} t} \ , \end{aligned} \tag{12}$$

(Here, $\mathbf{e_r}$, $\mathbf{e}_\theta$ and $\mathbf{e}_\phi$ denote the unit vectors in the r, θ and ϕ directions, and $\xi_r(r;n,l)$ and $\xi_h(r;n,l)$ radial and horizontal amplitude, respectively. The radial and horizontal amplitudes as well as the frequencies do not depend on m).

To be more specific, the unperturbed solution (12) is inserted into the rotational analogue to the boundary-value problem of Equation (2) [obtained from Eq. (11) in the same way as Eq.(2) follows from Eq. (1)]. To *first order* in $\boldsymbol{\Omega}$ one obtains, still in the co-rotating system (and with the asterix * denoting complex conjugation, and the indices of ξ henceforth dropped for convenience)

$$2\pi\Delta\nu_{nlm}^{(\text{co-rotating})} \ = \ i \ \frac{\int \xi^* \cdot \boldsymbol{\Omega} \wedge \xi \rho r^2 dr \sin\theta d\theta}{\int \xi^* \cdot \xi \rho r^2 dr \sin\theta d\theta} \ . \tag{13}$$

For the observer on earth, *i.e.* in the nonrotating frame (we disregard the earth's motion), the frequency of the modes with $m \neq 0$ are Doppler-shifted by the amount of $\pm m\Omega$ ($\Omega = |\boldsymbol{\Omega}|$, and the sign depends on the convention employed in defining prograde and retrograde modes). The result is (note that the ϕ component of ξ is proportional to m)

$$2\pi\Delta\nu_{nlm}^{(\text{nonrotating})} \ = \ m\Omega\,(1 - C_{nl}) \ , \tag{14}$$

with C_{nl} being the part due to Eq. (13). Expression (13) goes back to Cowling and Newing (1949) and Ledoux (1951) (the C_{nl} are sometimes called the Ledoux constants). In the case of the Sun, the C_{nl} are much smaller than 1 (typically by two orders of magnitude or more, see, *e.g.* Brown *et al.*, 1986, Gough, 1981), *i.e.* the term from the transformation back to the inertial frame (called advection term) is much more important than the dynamical effect of the Coriolis force on the mode.

To *second order* with respect to Ω, there are three different types of contributions to the frequency shifts. The first is the second-order perturbation term of the Coriolis force in Eq.(11), including second-order effect of the back-transformation to the nonrotating frame, the second is the dynamical effect on the modes and their frequencies of the centrifugal force in Eq.(11), and the third is the frequency change caused by the distortion of the solar cavity itself, *i.e.* by the change of the *equilibrium* solution due to the centrifugal force. All these second-order contributions can be shown to be small (typically at least one to two orders of magnitude below the linear advection term, see Dziembowski and Goode, 1984; Gough and Taylor, 1984). It turns out, therefore, that to a good approximation the advection term is sufficient to discuss the link between rotational splittings and the specific form of internal rotation.

4.1.2 Differential rotation

The generalization of the first-order term (14) to the case of nonuniform rotation was made by Hansen *et al.* (1977) and Gough (1981). Neglecting the Coriolis term (that would be a generalization of Eq. (13), with Ω becoming a function of depth and latitude θ), we are left with the advection term, which equally becomes a weighted integral of the angular velocity. The total first-order frequency splitting is, in the nonrotating observer frame (see *e.g.* Morrow, 1988a)

$$2\pi\Delta\nu_{nlm}^{(\text{nonrotating})} = -m\frac{\int \Omega(r,\theta)\xi^*\cdot\xi\rho r^2 dr \sin\theta d\theta}{\int \xi^*\cdot\xi\rho r^2 dr \sin\theta d\theta} \ . \tag{15}$$

4.2 *Inverse problem and results*

4.2.1 Internal rotation in equatorial plane

The rotational splitting (15) is symbolically written as (the angular part of the integral cancels)

$$2\pi\Delta\nu_{nlm}^{(\text{nonrotating})} = -m\int_0^R \Omega(r)K_{nl}(r)r^2 dr \ . \tag{16}$$

The functions K_{nl} are called *rotational kernels*, and they are simple functions of the unperturbed (*i.e.* nonrotating) eigenfunctions, *i.e.* the horizontal and vertical mode amplitudes [see Eq. (12)], which depend only on n and l. The specific form of the kernels is obtained by inserting Eq. (12) into Eq. (16). In the approximation chosen, all multiplets are strictly equidistant. Differential rotation as a function of depths is reflected by the *different spacing* of the multiplets, because obviously the high-l modes can only collect values of $\Omega(r)$ close to the solar surface, while the lower-l modes pick up the internal rotation deeper inside (see discussion of mode propagation, subsection 2.2). Therefore, the variation of the spacing of the multiplets considered as a function of l (and with a smaller influence also of n) should in principle reflect the course of the internal rotation of the Sun.

Before coming to the results obtained so far, I mention that a principal problem remains, of course, because there is only a finite set of data to yield the *function* Ω. Strictly speaking, such a problem is impossible to solve. Nevertheless, there are more or less intuitive procedures to invert the data. Numerical simulations with artificial data, *i.e.* with arbitrary given profiles of internal rotation, have been used to verify the feasibility of the procedure. In a typical procedure one assumes a piecewise constant angular velocity, *i.e.* $\Omega(r) = \Omega_i$ in the interval $[r_i, r_{i+1}]$. Such a step function for Ω reduces the integral equation (17) to a purely algebraic one. If the coarseness of the steps is suitably chosen for a real set of observed rotational splittings, one can arrange that a unique solution results, but one can always resort to even coarser step function and deal with the resulting over-determined problem using a least-square method.

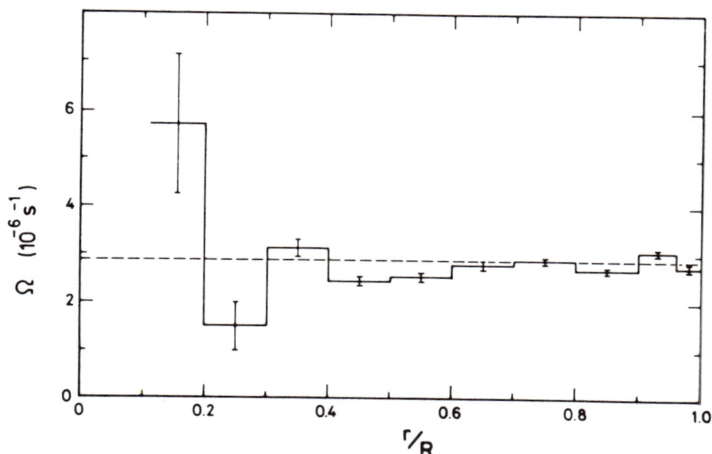

Figure 6. Estimated step function of the solar angular velocity in the equatorial plane, using the Duvall and Harvey (1984) data. The errors are computed from the estimated errors in the data; their magnitude also depends on the choice of the steps in Ω (from Duvall *et al.*, 1984).

The already 'classical' result obtained by Duvall *et al.* (1984) is shown in Figure 6. It reveals an angular velocity that is practically constant and equal to the surface equatorial value, except (perhaps) in the core region ($r < 0.3R$). There might be some evidence for a rapidly rotating core. Note, however, the large error bars there. They are essentially due to inadequate observational data, and not to limitations of the inversion methods. This is easily seen by considering that for a given mode with degree l the total width of the rotational multiplet is $\Delta\nu_{max} = 2l\nu_{l,rot}$, [where $2\pi\nu_{l,rot}$ is some average of $\Omega(r)$]. Since $\nu_{l,rot}$ is on the order of 0.46μHz (surface value), and since the observational precision of individual oscillation frequencies is also of the same order (see section 3) it is clear that low-order splittings, and thus the rotational velocity deep down, can only be very poorly determined (see *e.g.* Figure 5). For higher l, *i.e.* in the more shallow zones of the Sun, the observational situation becomes much better, since $\Delta\nu_{max}$ becomes much larger than the observational resolution. Compared to these observational difficulties, those stemming from the inversion method are unimportant. This is illustrated in Figure 7, which shows a typical result from a systematic study of numerical experiments (Christensen-Dalsgaard and Gough, 1984), in which artificial data (*i.e.* frequency multiplets) were generated for a whole series of arbitrary rotation profiles. Despite the theoretical limitations of such inversion (see above), there are no

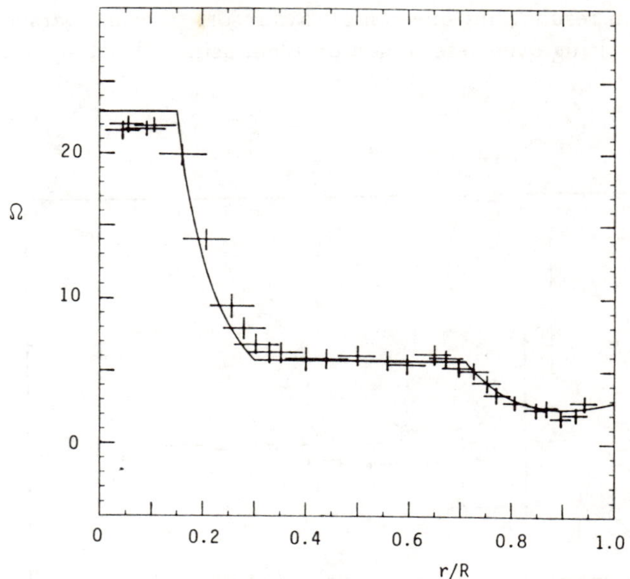

Figure 7. Typical success of numerical experiments performed to test the reliability of the inversion procedure of Figure 6. An arbitrary profile of internal rotation is given (solid line) and the oscillation frequencies and splittings of the modes analogous to ones used in the result of Figure 6 were computed. The crosses show the result of the inversion. The horizontal and vertical bars denote the estimated uncertainty of the inversion, obtained from a comparison of different methods (from Christensen–Dalsgaard and Gough, 1984).

problems with finding the assumed rotation law using the theoretically computed mode frequencies.

4.2.2 Latitudinal variation of internal rotation

As seen from Eq.(15), the nonequidistant parts of the frequency multiplets tell about the latitudinal variation of the internal rotation. Conveniently, the multiplets are developed in terms of Legendre polynomials (Eq. 9), and the resulting coefficients a_i become the relevant data. The determination of the coefficients a_1,a_3,a_5 is still rapidly improving, and systematic differences in the results from different groups occur (see Harvey, 1988). No definitive picture for the latitudinal variation has come up yet. Nevertheless, I show in Figure 8 one of the current interpretations by Morrow (1988a,b), which is consistent with the observations of Brown and Morrow (1987) (Figure 9). Other groups recently also have reported results and interpretations, differing from each other quite substantially, because of the not yet sufficient accuracy of the data (*e.g.* Korzennik *et al.*, 1988; Rhodes *et al.*, 1988; Tomczyk *et al.*, 1988; Christensen–Dalsgaard and Schou, 1988).

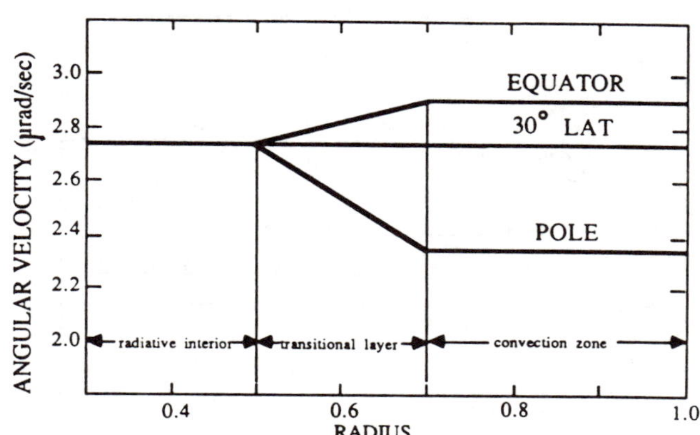

Figure 8. Morrow's (1988a,b) SRF model (surface-like through the convection zone) for internal rotation. The model is consistent with the data of Figure 9 (from Morrow, 1988a).

It is interesting to note that helioseismology has also discovered significant nonzero values for a_2 and a_4. Their values are much larger than the higher order terms of rotation and indicate the presence of asphericities not due to rotation [see, *e.g.*, and Kuhn (1988) and Libbrecht (1988), who in addition found a solar-cycle dependent variation of these even coefficients].

5. Conclusion

At the present speed of development it is only a matter of a few more years until our helioseismological picture of the solar internal rotation will become much clearer. Substantial experimental and theoretical effort is being made to overcome limitations in the form of day-night gaps, other symmetry-breaking effects on the Sun than rotation, and asphericity of the solar structure. The planned networks on earth (Birmingham group, SLOT, IRIS for low-angular resolution, GONG for full-disk imaging) will fill the observational gaps. The planned space instruments on board of SOHO (MDI, GOLF, VIRGO) will not only provide continuous coverage, but also avoid terrestrial noise.

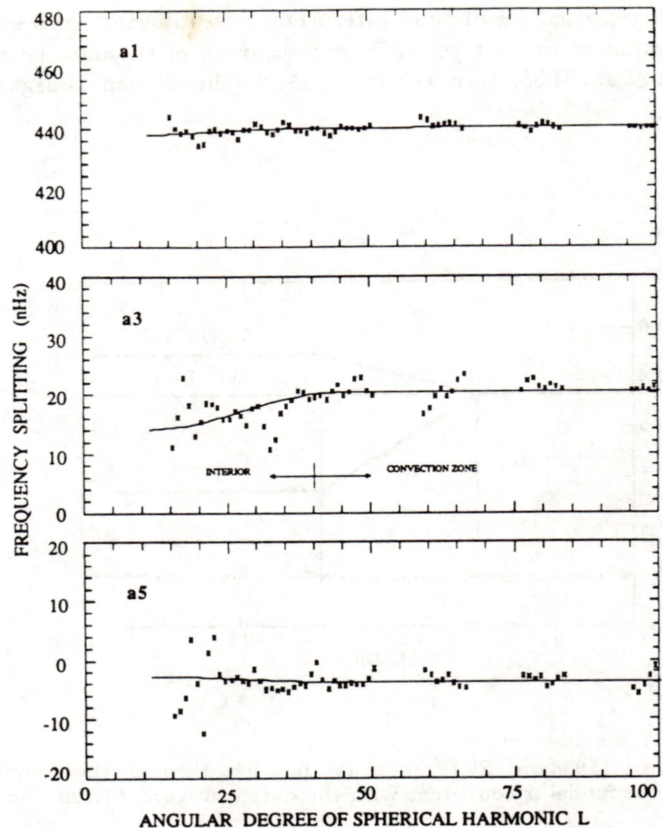

Figure 9. Coefficients a_1, a_3, a_5 of the Brown and Morrow (1987) data (dots). The solid lines correspond to the model in Figure 8 (from Morrow, 1988a).

6. References*

Brown, T.M. 1985, *Nature*, **317**, 591-594.

Brown, T.M., Mihalas, B.W., Rhodes, E.J., Jr. 1986, Solar Waves and Oscillations, in *Physics of the Sun, Vol. I*, ed. P.A. Sturrock (Reidel, Dordrecht), 177-247.

Brown, T.M., Morrow, C.A. 1987, *Astrophys. J.*, **314**, L21.

Christensen–Dalsgaard, J., Gough, D.O. 1984, in *Solar Seismology from Space*, eds. R.K. Ulrich *et al.* (Jet Propulsion Laboratory Publication 84-84), 79.

Christensen-Dalsgaard, J. 1986, in *Seismology of the Sun and the Distant Stars*, ed. D.O. Gough (NATO ASI Series, Reidel, Dordrecht), 3-18.

Christensen–Dalsgaard, J., Schou, J. 1988, Differential rotation in the solar interior, in TENERIFE, 149-153.

Christensen–Dalsgaard, J. 1989, Helioseismic measurement of the solar internal rotation, in *Proc. Annual Meeting of the Deutsche Astronomische Gesellschaft, Graz, 1989*, (Springer), (in press).

Christensen–Dalsgaard, J., Berthomieu, G. 1990, in *Solar Interior and Atmosphere*, eds. A.N. Cox, W.C. Livingston, & M. Matthews (Space Science Series, University of Arizona Press), in press.

Claverie, A., Isaak, G.R., McLeod, C.P., Van der Raay, H.B., Roca Cortés, T. 1979, *Nature*, **282**, 591-594.

Cox, A.N., Chitre, S.M., Frandsen, S., Kumar, P. 1990, in *Solar Interior and Atmosphere*, eds. A.N. Cox, W.C. Livingston, & M. Matthews (Space Science Series, Unversity of Arizona Press), in press.

Cowling, T.G., Newing, R.A. 1949, *Astrophys. J.*, **109**, 149-158.

Däppen, W., Dziembowski, W., Sienkiewicz, R. 1988, in *Advances in Helio- and Asteroseismology*, eds. J. Christensen–Dalsgaard and S. Frandsen (IAU Symp. 123, Reidel, Dordrecht), 233-247.

Deubner, F.-L. 1975, *Astron. Astrophys.*, **44**, 371.

Deubner, F.-L., Gough, D.O. 1984, *Ann. Rev. Astron. Astrophys.*, **22**, 593-619.

Duvall, T.L., Jr 1982, *Nature*, **300**, 242-243.

Duvall, T.L., Jr, Harvey, J.W. 1984, *Nature*, **310**, 19.

Duvall, T.L., Dziembowski, W.A., Goode, P.R., Gough, D.O., Harvey, J.W., Leibacher, J.W. 1984, *Nature*, **310**, 22.

Duvall, T.L., Jr, Harvey, J.W., Pomerantz, M.A. 1986, Latitude and depth variation of solar rotation, *Nature*, **321**, 500-501.

Dziembowski, W, Goode, P.R. 1984, *Mem. Soc. Astr. Italiana*, **55**, 185-213.

Gough, D.O. 1981, *Mon. Not. R. astr. Soc.*, **196**, 731.

Gough, D.O., Taylor, P.P. 1984, *Mem. Soc. Astr. Italiana*, **55**, 215-226.

Gough, D.O. 1985, in *Future Missions in Solar, Heliospheric and Space Plasma Physics*, ed. E. Rolfe, ESA SP-235, European Space Agency, Noordwijk), 183-197.

---

* I use the following abbreviated notation for the frequently cited Tenerife conference proceedings: TENERIFE = *Seismology of the Sun and Sun-like Stars*, ed. E.J. Rolfe (ESA Publications SP-286, Noordwijk, The Netherlands)

Hansen, C.J., Cox, J.P., VanHorn, H. 1977,The effects of differential rotation on the splitting of nonradial modes of stellar oscillation, *Astrophys. J.*, **217**, 151.

Harvey, J.W., Duvall, T.L., Jr 1984, in *Solar Seismology from Space*, eds. R.K. Ulrich *et al.* (Jet Propulsion Laboratory Publication 84-84), 165-172.

Harvey, J. 1988, Solar Internal Rotation from Helioseismology, in TENERIFE, 55-66.

Korzennik, S.G., Cacciani, A., Rhodes, E.J., Jr 1988, Inversion of the solar rotation rate versus depth and latitude, in TENERIFE, 117-124.

Kuhn, J.R. 1988, Radial and temporal structure of the internal solar asphericity, in TENERIFE, 87-90.

Lamb, H. 1932, *Hydrodynamics*, (Cambridge University Press).

Ledoux, P. 1951, *Astrophys. J.*, **144**, 373-384.

Leighton, R.B., Noyes, R.W., Simon, G.W. 1962, *Astrophys. J.*, **135**, 474.

Libbrecht, K.G. 1988, Solar p-mode frequency splittings, in TENERIFE, 131-136.

Libbrecht, K.G., Woodard, M., 1990, Observations of solar cycle variations in solar p-mode frequencies and splittings, in *Proc. Oji Seminar on the Progress of Seismology of the Sun and Stars*, eds. Y. Osaki and H. Shibahashi (Springer Lecture Notes), (in press).

Libbrecht, K.G., Woodard, M., Kaufman, J.M. 1990, *Astrophys. J. Suppl.*, (submitted).

Morrow, C.A. 1988a, *A new picture for the internal rotation of the Sun*, Ph.D. Thesis University of Colorado, Boulder (NCAR Cooperative Thesis No. 116, NCAR, Boulder, Colorado).

Morrow, C.A. 1988b, Solar rotation models and the a_1, a_3, a_5 splitting coefficients for solar acoustic oscillations, in TENERIFE, 91-98.

Newton, H.W., Nunn, M.L., 1951, *Mon. Not. R. astr. Soc.*, **111**, 413.

Pallé P.L., Pérez Hernandez, F., Régulo, C., Roca Cortés, T. 1988, Rotational splitting of low l solar p modes, in TENERIFE, 125-130.

Rhodes, E.J., Jr, Ulrich, R.K., Simon, G.W. 1977, *Astrophys. J.*, **218**, 901.

Rhodes, E.J., Jr, S., Cacciani, A., Korzennik, S.G., Tomczyk, S., Ulrich, R.K., Woodard, M.F. 1988, Radial and latitudinal gradients in the solar internal angular velocity, in TENERIFE, 141-147.

Schwarzschild, M. 1958, *Structure and Evolution of the Stars*, (Princeton University Press).

Tassoul, J.-L. 1978, *Theory of Rotating Stars*, (Princeton University Press).

Tomczyk, S., Cacciani, A., Korzennik, S.G., Rhodes, E.J., Jr, Ulrich, R.K. 1988, Measurement of the rotational frequency splitting of the solar five-minute oscillations from magneto-optical filter observations, in TENERIFE, 141-147.

Unno, W., Osaki, Y., Ando, H., Saio, H., Shibahashi, H. 1989, *Nonradial Oscillations of Stars (second edition)*, (University of Tokyo Press).

LOSS OF MASS AND ANGULAR MOMENTUM: THE OBSERVATIONAL POINT OF VIEW

Claude Catala
Observatoire de Paris, Section de Meudon
France

Summary of the review: The problem of mass loss and that of angular momentum loss from stars are closely coupled: when a star loses mass and rotates, it also loses angular momentum. It is therefore possible to derive information about angular momentum loss from studying mass loss phenomena (mass loss from T Tauri stars provides us with some insight into the past history of angular momentum of the Sun, for instance), and conversely, statistical studies of angular momentum loss can lead us to infer the existence of weak mass loss that would not be directly detectable (for instance, we know that solar-type stars lose mass because we have evidence that they lose angular momentum during their lives).

This review mainly focused on the observational point of view. A major part was devoted to describing the methods of mass loss detection and measurement:

P Cygni profiles: they constitute a direct evidence for outflow, and provide a convenient estimate of the maximum velocity reached in the region of formation of the lines considered. They can also yield estimates of mass loss rate, density law, velocity law, temperature law in the wind, through detailed modelling. A great emphasis was put on the problem of formation of P Cygni profiles, and on their diagnostic power. The problems linked to line profile interpretations were also discussed, as well as the uncertainties on the determination of the wind structure.

Line asymmetries: observed in chromospheric lines of red supergiants, they are interpreted as evidence for chromospheric outflows.

Infrared and radio continuum: they probe regions located further out in the stellar winds. Therefore, the determinations of mass loss rates from infrared or radio continuum observations do not have to rely on a precise modelling of the wind structure near the stellar surface.

Indirect evidence for mass loss: observations of rotation rates for cool stars of different ages provide us with an estimate of their rate of angular momentum loss, from which we can infer the presence of weak mass loss.

The rest of the review was devoted to a short survey of the mass loss phenomena across the HR diagram, from which it is apparent that all types of stars seem to lose mass at various rates. Examples:

High rates of mass loss (10^{-6}-10^{-5} $M_{\odot}yr^{-1}$) and high wind velocities ($\approx$ 3000 kms^{-1}) for WR stars and blue supergiants. These determinations are based on observations of P Cygni profiles and of radio continua.

High rates of mass loss ($\approx 10^{-6}$ $M_{\odot}yr^{-1}$) and low wind velocities ($\approx$ 10-50 kms^{-1}) for red giants and supergiants, inferred from observations of P Cygni profiles and line asymmetries.

Intermediate rates of mass loss ($\approx 10^{-7}$-10^{-8} $M_{\odot}yr^{-1}$) and intermediate wind velocities ($\approx$ 100-500 kms^{-1}) for pre-main sequence stars (T Tauri and Herbig Ae/Be stars), determined by detailed modelling of P Cygni profiles and observations of radio continuum in a few cases.

For a review on the subject, see :

Cassinelli, J. P., MacGregor, K. B., 1986, Physics of the Sun , P. A. Sturrock, T. E. Holzer, D. M. Mihalas, R. K. Ulrich, vol III, p. 47.

PRE-MAIN SEQUENCE EVOLUTIONARY TRACKS AND LITHIUM BURNING

I. Mazzitelli
Istituto di Astrofisica Spaziale
C.P. 67 - 00044 Frascati - Italy

Abstract. A review about the main physical uncertainties still weighing upon the computation of Pre Main Sequence (PMS) evolutionary tracks is given, in order to evaluate (at least qualitatively) their possible influences upon the surface stellar parameters. It is shown that we are not yet able to reliably compare observations with theoretical luminosities and colours, and also the correlation between initial stellar parameters (mass and chemical composition) and expected lithium depletion in PMS is not very sound. In particular, even if updated evolutionary models seem to show that Li-depletion in the Sun during PMS can be marginally consistent with the observations, several problems still remain unsolved.

In this framework, the observed spread of Li-abundances among Main Sequence (MS) stars belonging to a given young cluster, is attributed to the influence of rotation and magnetic fields (and perhaps accretion), which largely affect Li-depletion. The spread observed among different young clusters is instead likely to be due to intrinsic differences, mainly in the chemical compositions.

1. Introduction

The theoretical explanation of the observed abundances of lithium in PMS and young MS stars is, in spite of the appearent simplicity of the mechanisms involved, one of the most puzzling problems in stellar evolution. PMS stars should be in fact relatively well understood, at least in principle, thanks to their homogeneous chemical compositions, and to the lack of sharp physical profiles, after large-scale accretion and mass loss phases are over, and deuterium burning has lead the star to fully convective, hydrostatic conditions (Stahler 1988).

Theoretical expectances in the above framework are for a well defined relation between mass (or T_{eff}) of the star in MS and surface abundance of lithium, if we ignore (as in the following of this paper) possible mixing up to the surface, over long timescales, of Li-depleted matter coming from below the bottom of the convective envelope. This point has to be clarified since the beginning of the discussion. In fact, the possibility that mixing mechanisms acting during MS below the surface convective region can change, in billion years, the surface chemistry of solar type stars, is•to be taken seriously (Schatzman 1977). This is the reason why we limit our attention to very young MS stars, or PMS stars only.

Instead of the clean framework depicted by theory, observations show a much more chaotic behaviour, where not only different clusters show different relations between lithium and T_{eff}, but also stars belonging to the same cluster show a wide dispertion of values of surface Li-abundances for the same value of T_{eff} (see for instance Balachadran et Al. 1988, Stauffer et Al. 1989, Strom et Al 1989).

Actually, as we will see in brief, several physical mechanisms not yet well understood, having little or no effect upon the structure of MS stars, are instead much more relevant in PMS phases, particularly when the star is still close to the Hayashi track (Mazzitelli 1989), so that the supposed simplicity of PMS stars is more a trivia than a reality. Not all these sources of uncertainty have the same effect upon the Li-destruction; some of them happen to affect only other surface parameters of stars, but also in these cases they have some relevance for the Li-problem, since they influence at least the relations between Li-depletion in PMS and observational signatures.

The main pourpose of the following discussion will be that of recalling the main points in which our present understanding of the theory of stellar evolution is insufficient to give realistic estimates about the structures of PMS stars. The effects of each of the more evident uncertainties in determining the surface parameters (Luminosity and T_{eff} versus mass and age) and the central ones (temperature and depth of convection) will be considered separately, to avoid confusion. In fact, it is worth stressing since the beginning that two physical uncertainties having nearly the same effect upon the position of a PMS star in the HR diagram, do not have necessarily the same effect also upon Li-destruction in the interior. Some effects, like for instance the uncertainties in the overadiabatic convection theory, have influence mainly in determining the surface temperature of the star, whyle their effect upon the central temperature, or on the deepening of convection down to the center of the star, is negligible. Other effects, like the uncertainties in the radiative opacity at intermediate temperatures, affect both the surface and the central regions.

2. General concepts

PMS stars are, for a fraction of their lifes, fully convective from the center to the base of the optical atmosphere. At a given point of their evolution, stars having masses larger than about 0.40 $M_{\odot}$ (at least for a Pop I chemical composition, D'Antona and Mazzitelli 1982) begin to develope a radiative core, and move away from the Hayashi track. The larger is the star's mass, the lower is the central temperature when the radiative core appears (D'Antona and Mazzitelli 1984, thereafter DM), and the earlier is the decoupling between central and surface chemical composition. The computations show in fact that complete depletion of a given element at the surface, is possible only if the element is burning at the very center of a fully convective star. Burning at the base of a convective envelope, even a very deep one, gives rise to small or negligible depletion at the surface since:

-the density (and the reaction rates) fastly decline by orders of magnitude when moving from the center towards the surface, and:

-as soon as convections begins receding from the center, the temperature at the base of convection decreases in a timescale shorter than the one for nuclear burning.

Mainly for these reasons, the larger is the star's mass, the less the star is able to deplete fragile elements at the surface. Also the fact that "convective undershooting" (DM, VandenBerg et Al. 1989) is able to provide larger depletions at the surface for a given star's mass, is to be seen in this same framework, that is: undershooting works in the direction of maintaining chemically mixed the core with the surface for a longer time, but as soon as undershooting moves away from the center, it has no relevance any more.

Given that, it is clear that any physical mechanism affecting convection could, in principle, influence also the relation between star's mass and depletion of fragile elements at the surface in PMS. Actually, as already recalled, some mechanisms affect only the surface parameters of the star; others play a role both for what concerns surface and central convection, so that they influence also the nuclear depletion at the surface, while a last class of mechanisms has some effects on the mixing of the core only, leaving unchanged both luminosity and T_{eff}, but leading to larger or lower depletions of fragile elements at the surface. In the following, I will try to quantitatively discuss overadiabatic convection, thermodynamics, radiative opacity and convective undershooting, and will also try to raise at least qualitative conclusions for what concerns rotation, and magnetic fields.

3. The overadiabatic convection

Convection in stars cannot be strictly adiabatic, since otherways it could not transfer energy. On the other hand, as soon as the local density of matter begins to be larger than about 10^{-2} g cm^{-3}, it is easy to show that the difference between the actual convective gradient and the adiabatic gradient (just the "overadiabaticity") required to carry away all the energy flux, is so small that it can be conveniently neglected in the computations. Unfortunately, PMS stars are not only dominated by convection, but they are also expanded, low density configurations. A reliable theory of overadiabaticity would then be required for making sound predictions about the behaviour of PMS stars, but the present status of the theory is not encouraging at all.

The commonly adopted schematization for the overadiabatic convection theory in all the stellar evolution simulations is, in fact, only a minor updating (Cox and Giuli 1968) of the former "Mixing Length" (ML) theory by Bohm-Vitense (1958), which is in turn only a <u>parametric</u> theory. The basic underlying physics in the ML theory is that convection in stars is expected to be largely turbulent, since any value for the Reynolds number one can derive for any stellar condition (and also allowing for the fact that a star has no definite walls, so that the characteristic volume is of the order of magnitude of the star's volume itself) is several orders of magnitude larger than the value expected to give rise to turbulence in a three-dimensional fluid. The ML theory assumes that all the energy is transferred by the "bubbles" having energies corresponding to the maximum of the spectral distribution (a delta function, see Canuto 1990), completely ignoring any contribution both from the growth and the Heisenberg-Kolmogoroff regions.

Since it is not to be expected a priori that souch a rough treatment could give rise automatically to a reliable match with the observations, a free parameter is then introduced, to be fixed in such a way that the observed surface properties of the Sun (which is presently the only star for which we have enough physical and chemical informations) are fitted by theory. This parameter is just the ML, defined as the size of the average convective bubbles, usually expressed in fractions of the Pressure Scale Heigth (H_p)

In principle, the theory could work, and all the successes of the present stellar evolution simulations are indeed grounded upon such a not very firm basement, even if at least two main objection can be raised against the ML theory, that is:

-there is no reason at all why a tuning performed upon the physical conditions prevailing in the Sun should hold also in other, completely different stellar conditions (from the red supergiants to the brown dwarfs), and:

-in any case, the values of ML required to fit the Sun are so large (ratios ML/H_p about 1.5 - 2) that it is hard to believe that the bubbles can maintain spherical shapes (which is the hypotesis when computing the radiative losses from the bubbles themselves), no to say that such large bubbles are not consistent with the hypotesis of strong turbulence.

If we try to rise some informations from observations of other stars, we have no firm conclusions, but the best we can say is that, if the value of ML required to fit the Sun gives a reasonable fit for all the MS band, the shape of the red giant branches both for open and globular clusters cannot be presently matched, if we do not allow for a systematic decrease of the value of the ratio ML/H_p when moving towards the red of the HR diagram (Chieffi and Straniero 1989), at least if we believe the correlations among colours, T_{eff} and bolometric corrections.

Coming back to PMS, and waiting for a better theory of overadiabatic convection, what we can presently do is to check if and how much a variation in the value of the ratio ML/H_p along the PMS evolution can modify both the surface and the central parameters of the stars. According to the above quoted observations, the tests have to be performed by using a lower value of ML/H_p along the Hayashi track than in MS.

A detailed discussion of the effects of changing ML/H_p in PMS can be found in Mazzitelli (1989) for what concerns the surface conditions, and in DM as far as Li-burning is concerned. The main conclusions can be summarized as follows:

-the surface conditions are dramatically affected by even a very small decrease in ML (from 1.6 to 1.5 H_p), so that all the comparisons between observations of PMS stars and evolutionary tracks cannot ignore this uncertainty. Roughly speaking, a PMS star for which the observed magnitudes and colours would correspond to a mass of 0.8 $M_\odot$ when compared to constant-ML theoretical tracks, can easily match the tracks relative to 0.9-1.0 $M_\odot$, if we allow for a slight variation of ML along the evolution;

-the effect of a changing in the value of the ratio ML/H_p upon the Li-depletion at the surface, is instead very small. Overadiabatic convection has a large influence only upon the very external layers of the star, and the temporal behaviour of the central temperature and the deepening of convection are not affected.

4. Thermodynamics

In cool stars, the gas does never behave as an ideal gas, not even in low-density structures as PMS stars. Due to the relatively low kinetic energies of the particles, in fact, the configurational effects (mainly those due to surrounding ions) can play a not negligible role in ionization. In the cases of expanded envelopes (PMS and Red Giants), hydrogen ionization is not particularly affected, but helium ionization, which takes place in inner regions and in an environment dominated by already ionized hydrogen, is somewhat more sensitive to the real-gas effects.

The presence of ionized hydrogen tends in fact to slightly increase the degree of ionization of helium with respect to what one should expect according to the Saha formula (Fontaine et Al. 1977, Magni and Mazzitelli 1979). The ionization region of helium in the density-temperature plane is then wider than for an ideal gas, and this has an effect also upon the values of the adiabatic temperature gradient in that region, since ionization causes a decrease in the adiabatic exponent γ (and in the adiabatic gradient) which is more marked, the sharper is ionization itself. If ionization becomes less sudden, also the decrease in the values of the adiabatic gradient is less peaked. In practice, for an ideal gas, one obtains in the ionization regions values of the adiabatic temperature gradient as low as 0.07-0.08, whereas for a real gas it is very hard to reach values below 0.10-0.12.

Also this effect upon PMS evolutionary tracks has been discussed by Mazzitelli (1989). A larger adiabatic gradient causes also a larger drop between the central and surface temperature of a PMS star, so that the tracks computed with real-gas equations of state are systematically cooler than those computed according to the Saha ionization equilibria. We still do not have sufficiently good thermodynamic treatments as to precisely quantify this effect, but we expect it to be of the same order of magnitude as the above discussed effect of the ML, so that it cannot be ignored in the observational vs. theoretical comparisons.

As for Li-burning, uncertainties in thermodynamics should not give rise to large differences. In fact, also if He-ionization takes place in a much deeper region than the one interested by overadiabaticity, nevertheless the ionization temperature of helium (10^5 K) is so lower than the Li-burning temperature that the effect of a non-ideal gas treatment upon Li-depletion should in any case be of minor interest, even if no detailed theoretical comparisons have yet been performed.

5. The radiative opacity

In a zero-order approach, the influence of radiative opacity upon a fully convective structure should be almost negligible. As a matter of fact, opacity turns out to be the most relevant parameter in determining both the surface and the central conditions of PMS stars. In fact, radiative opacity has the two main effects of:

-determining the value of the overadiabatic gradient in the convective external layers, through the value of the radiative gradient, and:

-determining the temperature at which the center of the star becomes radiative.

Of course, in the first case, what really matters is the low-temperature opacity, dominated by molecules, hydrogen and helium ionizations, and by the first ionizations of metals. In the second case, since the temperature is of the order of 10^6 K or larger, the main sources of opacity come from free electrons, and from the ionization of the inner shells of metals. It is worth recalling the present status of the art about the computations of radiative opacities.

Apart from very old semi-analytical formulations, the first opacity tables of wide use in stellar modeling were the ones by Cox and Stewart (1970), supplemented by the ones by Cox and Tabor (1976) for other chemical compositions. Those tables were supposed to cover the whole range of physical conditions of evolutionary interest, but did not account for most molecules at low-temperature, and for inner ionizations of metals at large temperatures, so that they were probably underestimated.

Later on, Huebner et Al. (1977) provided the so called "Los Alamos Opacities", accounting for the inner ionizations of most metals. These opacities were provided separately for each chemical element and, in order to get the final Rosseland mean opacity for a given chemical composition, one had to merge all the tables. Of course, this procedure does not allow for evaluating the influence of each single element upon the opacity of another element due, for instance to the formation of molecules or to the capture of a free electron. To avoid misuse of these opacities, then, the tables were provided only for temperatures larger than 10^4 K, where molecules should be of no interest. Consistently with the expectances, the new opacity values were found to be larger than the previous ones, but their use was limited by the fact that, in order to compute full stellar models, one had to supplement them with the old low-temperature opacities below 10^4 K.

A first, serious attempt to include in the low-temperature opacities also molecules, was performed by Alexander (1975). Unfortunately, it turned out that the molecular opacities, particularly those for the water vapour, were largely overestimated, due to intrinsic difficulties in treating a very complex physical framework, so that those opacities were soon abandoned.

Later on, Alexander et Al. (1983) revised the opacities downward and, very recently, Alexander (1989) has also solved the problem of H_2O opacity. He is now computing low temperature opacities including molecules, but no tuning of these opacities upon evolutionary models still exists.

In practice, the present situation is that we can make use of probably reliable opacities for the internal layers of the stars, but the opacities for the surface layers are still in run. Let us try to understand what this can mean for PMS stars.

An underestimate of the low temperature opacities will give rise to lower than real overadiabatic temperature gradients, and lower then real differences in temperature between center and surface of the star. In other words, new low temperature opacities including molecules will tend to give rise to PMS evolutionary tracks having T_{eff} lower than the present ones (see Mazzitelli 1989). Again, this has to be considered when comparing observations and theoretical tracks.

As for the high temperature opacities, since they are larger than the previous ones, it is to be expected that in a star of given mass, surface convection will stick to the center for somewhat longer, and up to a larger central temperature than previously estimated. For a given mass, then, the new opacities will give rise to larger surface depletions of light elements than the former ones. This is just what is found when comparing the evolutionary tracks by DM and those by VandenBerg (1989) for the 1.0 $M_{\odot}$ star.

The above tracks, in fact, differ mainly in the high temperature opacities. In the updated computations by VandenBerg, a significant Li-depletion at the surface of 1 $M_{\odot}$ star is found with an extra-mixing of only 0.25 H_P, whereas in DM an extra-mixing three times larger was required to fit the Sun. We can then say that, by now, the observed Li-depletion in the Sun is not far from being explained by plain PMS depletion, and that a new updating of the high temperature opacities, which will probably result in a further increase in their values because of the inclusion of further atomic transitions, will probably explain the observed Li-abundance in the Sun without further hypoteses, like for instance extra-mixing. Most unfortunately, when we will discuss magnetic field, we will see that this is not yet the real solution of the problem.

That the high temperature opacities were by far the dominant mechanism in determining the Li-depletion at the surface, was already clear in the results by DM. In fact, it was there shown that, while no significant Li-depletion was expected without extra-mixing for a 1 $M_{\odot}$ star, an increase of 50% in the opacities (mimicked by an increase in metal abundance from Z=0.02 to Z=0.03) was sufficient to give rise to substantial Li-depletion even for an 1.2 $M_{\odot}$ star.

It can be useful to summarize at this point the range of results one can expect for a theoretical evolutionary track upon the H-R diagram, if we allow overadiabaticity, thermodynamics and low-T radiative opacities to vary within relatively strict, perfectly acceptable boundaries, in view of the basic theoretical uncertainties.

Figure 1 shows three evolutionary tracks in PMS. The continuous lines are relative to the evolutions of Pop I stars of, respectively, 0.7 and 1.0 $M_\odot$, in the classical evolutionary framework, that is:

-constant ratio $ML/H_p = 1.6$;

-Saha equilibria in the thermodynamics, and

-Cox and Stewart (1970) radiative low-T opacities.

The dashed line is instead relative to a 1 $M_\odot$ star of the same chemistry, but computed according to the following recipes:

-variable ratio ML/H_p, tuned in such a way that it linearly changes with Log T_{eff}. It reaches a maximum value of 1.6 at Log T_{eff}=3.70, and decreases to 1.5 at Log T_{eff}=3.60.

-"intermediate equation of state", in the sense that the values of the adiabatic gradients are weighted averages between the ones evaluated according to the Saha ionization equilibria, and the ones coming from the equation of state by Magni and Mazzitelli (1979). The relative weights are 2 for the Saha thermodynamics, and 1 for the Magni and Mazzitelli equation of state.

-increased values for the low-T opacities. Opacities linearly increase when decreasing T, in such a way that they do not change at all at T = 10^4 K, and they are 30% larger than those by Cox and Stewart at T = 3000 K.

This last track is defined the "best guess" (b.g.) track, since it has been tuned according to the only reliable determination of mass for a PMS star. In fact, we have for the T Tauri star and X-ray source 162814-2427 in the Rho Ophiuchi dark cloud, a dynamical lower limit for the mass of the primary (the star is a two spectra spectroscopic binary) of about 1.0 $M_\odot$ (Mathieu et Al. 1989), whereas the theoretical PMS tracks would give about 0.7 $M_\odot$. The b.g. track shows that consistency between theory and observations is well within the theoretical error box.

As for Li-depletion at the surface, it turns out that the b.g. track exibiths exactly the same depletion (negligible) as the 1 $M_\odot$ track computed according to the standard input physics. This is consistent with the expectances, since no variation in the intermediate and high-T opacities had been introduced in the computations. The author recalls the attention of the readers

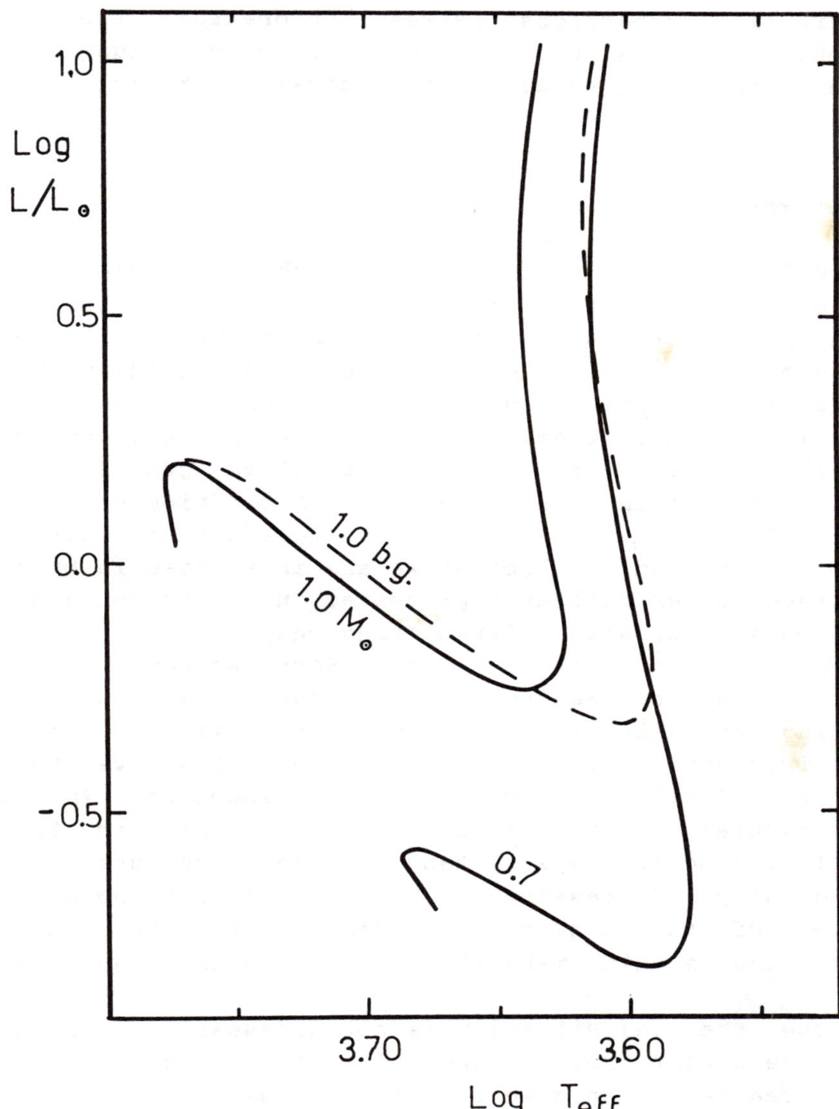

Figure 1: the PMS evolutionary tracks for Pop I stars of 1.0 and 0.7 $M_\odot$, with standard input physics (solid lines), and a "best guess" (b.g.) track for 1 $M_\odot$, always with "reasonably standard" input physics (dashed line), which gives the order of magnitude of the present theoretical error box in the H-R diagram.

upon Figure 1: comparisons between theoretical tracks and observational colours and luminosities cannot ignore the possible existence of <u>very large</u> theoretical uncertainties.

6. The extra-mixing

Let us define "extra-mixing" all the mechanisms which can be responsible for chemical mixing of matter, other than convection. Possible mechanisms are for instance convective over/undershooting, semiconvection, meridional circulation and so on. For the radiative regions of PMS stars we can exclude semiconvection and meridional circulation, since semiconvection requires the existence of chemical composition gradients, and meridional circulation is expected to work on timescales much longer than the PMS evolutionary times. We are then left with over/undershooting and, as long as we are interested in bringing to the surface matter nuclearly processed in layers beneath the bottom of convection, with undershooting only.

It is only fair to admit that, at present, we have no idea at all about the physical mechanisms which can be responsible for undershooting. Actually, in a star, convection is not simmetric and, in a first approximation, one could think that overshooting is much more likely to occurr than undershooting. In fact, convective bubbles coming from inside, when reaching the top of the formally convective layers, tend to break down into a region of lower density and pressure, whereas bubbles sinking down to the bottom of the convective region meet the opposite conditions, and should make their way in denser and denser layers.

Of course, the real situation is not necessarily so simple and neat; one cannot, for instance, forget that a non-linear, anelastic treatment can give rise to not negligible downward penetrative convection (Massaguer et Al, 1984), distinct from plain "undershooting", at least within a simplified physical scenario. However, when recalling the fact that, at present, there is still considerable dispute about the very existence of overshooting itself, it is clear that in the best of the cases we can treat undershooting just as a useful work hypotesis ad hoc tuned, to gain informations about how much our theoretical predictions agree with the observations, and how slowly or fastly our theoretical framework is changing.

Luckily enough, undershooting in PMS is not expected to influence the physical structure of the star, so that its only effect is to keep the surface chemistry sticking to the central one, and we are allowed to numerically experimenting with the models, without having to deal with secondary, tricking consequences upon luminosity or T_{eff}.

As already quoted, the recent computations by VandenBerg (1989) show that, in order to get relatively large Li-depletions in the Sun during PMS, only 0.25 H_P of undershooting are required, whereas in the former computations by DM, the required amount of undershooting was about 0.7 H_P. These results are conforting, since it appears that theory is approaching observations without ad hoc hypoteses. Actually, in the author's opinion, one will be allowed to claim that PMS Li-depletion explains the present Li-abundance in the Sun, only if and when theoretical models requiring <u>no undershooting at all</u> will predict substantial PMS Li-depletion for stars somewhat larger than the Sun, also in view of the following discussion about magnetic fields.

7. The rotation

With rotation, we are fastly approaching (and somewhat... overshooting...) the boundaries of the present generation stellar modeling, at least if we want our evolutionary results to be of general validity, and not restricted to peculiar objects. After all, a fraction of the contributions to the present volume indicate the extreme physical complexity of the whole framework.

In principle, rotation could be discussed together with magnetic fields, since the interplay between these two mechanisms is perhaps more significant than either of the two, in the resulting chemical mixing. Since however we <u>do</u> have some evolutionary result about rotation, whereas the results about magnetic fields are more <u>structural</u> than <u>evolutionary</u>, let me treat in this section rotation only.

In the stellar modeling, the inclusion of rotation is not mainly a matter of numerical problems or time-consuming computations; in fact, on the one side we have by now access to sufficiently large computing facilities, and on the other side the schematization by Kippenhahn and Thomas (1970), which does not require truly bi-dimensional computations, is more than adequate for the most of the cases of evolutionary interest. The real problem is that we do not yet know how to deal with the spatial and temporal evolutions of rotational angular momenta in stars.

In fact, during the evolution, viscosity, convection, compression and expansion, magnetic fields and so on are responsible for internal readjustments of the rotation rates, whereas the temporal evolution of the total angular momentum of the star is a complex function of mass loss and magnetic fields. At present, we can only try to get semiquantitative results, by making ad hoc (and "reasonable") hypoteses about all these mechanisms.

From the point of view of the surface parameters of PMS stars, rotation tends to decrease the surface gravity, so that the star mimicks the surface conditions of a smaller mass star, i.e.: shows a lower value of T_{eff}. This is precisely what is found in the models by Pinsonneault et Al. (1989); a 1 $M_{\odot}$ PMS evolutionary track computed by including in the code rotation, is roughly similar to the PMS track one would expect for a 0.95 $M_{\odot}$ star without rotation.

The effect of rotation upon Li-depletion is instead still completely unknown. In principle, rotation should help developing instabilities leading to mixing of some kinds, favouring larger Li-depletions at the surface, but this is only a qualitative point of view. Much, much work still remains in this framework before reaching firm conclusions.

8. The magnetic fields

As for the magnetic fields, when they are included in the treatment of overadiabatic convection (Moss 1968), they give rise to larger values of the overadiabatic temperature gradients in the external layers, leading again to cooler surface temperatures. At present, this is the only semi-quantitative argument which can be included in fully evolutionary computations.

When discussing the expected effects of magnetic fields upon the central conditions, at least for their influence upon the surface Li-depletion, we are probably touching the very heart of the problem. Qualitatively, we can say that, the stabilizing effect of the magnetic fields upon any matter movement should inhibit not only any kind of extra-mixing, but even cause an earlier detachment of convection from the center. Magnetic fields are then expected to act as a powerful mechanism preventing Li-depletion at the surface.

We do not have at present any reliable computation for PMS conditions; perhaps, some arguments can be found in computations relative to the Sun, for which some preliminary results already exist. In this framework, according to Spruit (1990), the effect of even a very small magnetic field upon solar turbulence should be so severe, to completely inhibit mixing below the surface convective region. If the same conclusions have to be applied also to PMS stars, in which the presence of large convective regions could be expected to give rise to powerful dinamo mechanisms, we have to face with the fact that magnetic fields, which we presently cannot treat in fully evolutionary frameworks, are by far the dominant mechanism in determining the degree of Li-depletion at the surface in PMS (but probably also in MS) phases. Given that, the immediately following conclusion is that the stellar masses which, for a given chemical composition, show surface Li-depletion in PMS, are only upper limits, and that the real values of limiting mass for PMS Li-depletion in PMS, in the presence of magnetic fields, can be even of several tenths of $M_{\odot}$ lower.

9. Conclusions

When summing up all the above conclusions, several uncertainties, but also some relatively firm statements, appear to arise. The first statement is for instance relative to the Sun or, better, to stars of approximately the same mass of the Sun. It is clear that stars about 1 $M_{\odot}$ are at the boundary of the Li-depletion/non-depletion in PMS. It is just among solar type stars that one can expect to observe the most of the spread in surface Li-abundances, since slightly different chemical compositions, or rotational and magnetic histories, make orders-of-magnitude differences in PMS Li-depletion.

Another statement is that, not only our present theoretical computations of PMS stars surface conditions are relatively uncertain, but also that surface conditions themselves (mainly the relation between stellar mass and T_{eff}) are a strong function of the chemical parameters, through the surface radiative opacities. Any comparison between theory and observations must allow for the different chemistries of the different clusters, and the comparison of all the observed clusters to evolutionary tracks of a unique chemical composition is likely to provide artificial and unnecessary spreads among the results.

In practice, while stars belonging to the same cluster, and having then the same initial chemical composition, have their Li-depletion at the surface determined by intrinsic differences in their rotational, magnetic and, probably, accretion histories, so that it is legitimate to expect some spread in the Li vs. mass relation for a given cluster, care has to be taken in comparisons among different clusters. It is in fact likely that at least some of the spread observed in the Li vs. mass relations among different clusters be due to an incorret evaluation of mass, due in turn to the use of evolutionary tracks not suitable for the chemical composition of the cluster.

References

Alexander, D.R.: 1975, Astrophys. J. Suppl. 29, 363.
Alexander, D.R.: 1989, (private communication).
Alexander, D.R., Johnson, H.R., Rypma, R.L.: 1983, Astrophys. J. 272, 773.
Balachandran, S., Lambert, D.L., Stauffer, J.R.: 1988, Astrophys. J. 333, 267.
Bohm-Vitense, E.: 1958, Z. fur Astrophys. 46, 108.
Canuto, V.M.: 1990, Astron. Astrophys. (in press).
Chieffi, A., Straniero, O.: 1989, Astrophys. J. Suppl. 71, 47.
Cox, A.N., Stewart, J.N.: 1970, Astrophys. J. Suppl. 19, 243.
Cox, A.N., Tabor, J.E.: 1976, Astropys. J. Suppl. 31, 271.
Cox, J.P. and Giuli, R.T.: 1968, Principles of Stellar Structure, Gordon and Breach, New York.
D'Antona, F., Mazzitelli, I.: 1982, Astrophys. J. 260, 722.
D'Antona, F., Mazzitelli, I.: 1984, Astron. Astrophys. 138, 431. (DM)
Fontaine, G., Graboske, H:E:, Van Horn, H.M.: 1977, Astrophys. J. Suppl. 35, 293.
Huebner, W.F., Marts, A.L., Magee, N.H., Argo, M.F.: 1977, Los Alamos Sci. Lab. Rept. (LA-6760-M).
Kippenhahn, R., Thomas, H.C.: 1970, "Stellar Rotation", ed. A. Slettbak (Dordrecht: Reidel), p. 20.
Magni, G., Mazzitelli, I.: 1979, Astron. Astrophys. 72, 134.
Massaguer, J.M., Latour, J., Toomre, J., Zahn, J.P.: 1984, Astron. Astrophys, 140, 1.

Mathieu, R.D., Walter, F.M., Myers, F.C.: 1989, Astron. J. 98 (3), 987.

Mazzitelli, I.: 1989, "Proceedings of the ESO workshop on star formation", in press.

Moss, D.L.: 1968, Monthly Not. Roy. Astron. Soc. 141, 165.

Pinsonneault, M.H., Kawaler, S.D., Sofia, S., Demarque, P.: 1989, Astrophys. J. 338, 424.

Schatzman, E.: 1977, Astron. Astrophys. 56, 211.

Spruit, H.: 1990 (this volume).

Stahler, S.W.: 1988, Astrophys. J. 332, 804.

Stauffer, J.R., Hartmann, L.W., Jones, B.F., McNamara, B.R.: 1989, Astrophys. J. 342, 285.

Strom, K.M., Wilkin, F.P., Strom, S.E.: 1989, Astron. J. 98 (4), 1444.

VandenBerg, D.A. and Poll, H.E.: 1989, Astron. J. 98 (4), 1451.

OUR PHYSICAL KNOWLEDGE

MASS TRANSPORT AND MIXING BY WAVES

Edgar Knobloch
Department of Physics, University of California,
Berkeley CA 94720

Both the solar neutrino flux and the abundances of the light elements, particularly Lithium, are very sensitive to mixing in the interior of the Sun. Such mixing processes may be intermittent and catastrophic as in the ^3H_e-driven instability suggested by Dilke and Gough (1972), or they may be continuous but gentle, a result of the small amplitude waves that are expected to be supported by the radiative interior. Evry Schatzman, in particular, has explored in detail the consequences of such weak but continual mixing, finding that a wide variety of observational facts falls into place if mixing characterized by a turbulent Reynolds number R_e of order 100 is assumed to be present, as reviewed elsewhere in this volume. The origin of this mixing is left unspecified, however. It was this problem that largely motivated my work with Henk Spruit on instabilities in the radiative core. Indeed there are a number of possible mechanisms which could lead to mixing. These include the baroclinic instability, a dynamical instability largely confined to equipotential surfaces (Spruit and Knobloch, 1984). A variety of doubly- and triply-diffusive instabilities may also occur, but these are again largely confined to equipotential surfaces (Knobloch and Spruit, 1983), as are the shear instabilities (e.g. Zahn, 1987). In addition to these mostly wavelike instabilities, the solar interior can also support a variety of neutrally stable waves. These include the g-modes, perhaps excited by the ^3H_e- instability in the core (Merryfield *et al*, 1989) , or by the convective overshoot (Press, 1981; Hurlburt et al, 1986), the well-documented p-modes, as well as a variety of wavelike modes associated with any magnetic fields that may be present.

It is clear that the problem of explaining the "observed" levels of mixing divides into two parts. The first requires an understanding of the wave generation mechanism, either an instability or a source. In the former case it is necessary to determine not just the waveform but also the amplitude at which it saturates, and hence nonlinear theory is required. In the latter it is essential to understand the spatial and temporal spectrum of the perturbations, as well as their amplitudes, that act as the source for the excitation of the waves, and the evolution of the wave, particularly its amplitude, as it propagates away from the source. For the purposes of discussing energy, momentum or angular momentum transport by waves it is in addition necessary to understand the way the waves deposit these quantities. This requires an understanding of wave absorption, reflection and of wave-breaking. The second part requires a study of the particle motion in the wave to determine whether any mixing takes place. This part naturally concentrates on the Lagrangian motion of the particles, and is inherently nonlinear, since the particle motion obeys the equation

$$\dot{\mathbf{x}} = \mathbf{u}(\mathbf{x}, t) \quad , \tag{1}$$

where $\mathbf{u}$ is the velocity field of the wave. This review focuses on this latter aspect of the problem. I discuss in some detail some of the considerations that must go into any believable theory of mixing. The questions asked, " when do waves transport mass", " what is meant by diffusion by waves", "does a diffusion coefficient exist", are very basic, but because of the nonlinear nature of (1), have perhaps unexpected answers.

It is important to distinguish between the **kinematic** problem in which $\mathbf{u}(x,t)$ is specified a priori, and the more difficult **dynamical** problem in which $\mathbf{u}(\mathbf{x},t)$ is sought as a solution of the nonlinear equations of motion of the fluid. An example of the former is given by

$$\mathbf{u}(\mathbf{x}, t) = \Re_e \left(\int \mathbf{A}(\mathbf{k})\, e^{i\omega(\mathbf{k})t - i\mathbf{k}.\mathbf{x}}\, d^3k \right) \quad , \tag{2a}$$

i.e. a superposition of disturbances with dispersion relation $\omega(\mathbf{k})$, and amplitude distribution $\mathbf{A}(\mathbf{k})$. With $\mathbf{A}(\mathbf{k})$ infinitesimal (2a) is a solution to the linearized equations of motion.

To appreciate the consequences when $\mathbf{A}(\mathbf{k})$ is finite, consider the one-dimensional monochromatic wave train (a sound wave),

$$u(x,t) = A \cos(kx - \omega t) \quad , \tag{2b}$$

travelling with the phase speed $c \equiv \omega/k$. To solve for the particle motion, one expands x in powers of the amplitude A :

$$x = x_0 + A\, x_1 + \ .... \quad . \tag{3}$$

Then

$$\dot{x} = A \cos(kx_0 - \omega t) + \frac{A^2}{c} \sin^2(kx_0 - \omega t) + ... \quad . \tag{4}$$

Hence at leading order in A the particle simply oscillates about its initial position x_0. At second order, however, one finds that the particle drifts in the direction of the phase velocity of the wave, with a speed

$$< \dot{x} > = \frac{1}{2}\frac{A^2}{c} + O(A^4) \quad , \tag{5}$$

where

$$< ... > = \frac{2\pi}{\omega} \int_0^{\frac{2\pi}{\omega}} (...)\, dt \quad .$$

This is known as Stokes' drift and is a purely nonlinear effect. In fact it is possible to calculate $< \dot{x} >$ from (2b) exactly (Knobloch and Weiss, 1987a),

$$< \dot{x} > = \begin{cases} c\,[1 - \sqrt{1 - (A/c)^2}] & , \ A \le c \\ c & , \ A \ge c \end{cases} \tag{6}$$

showing that the particles are trapped in the wave and therefore carried with its phase velocity, if the wave amplitude is sufficiently large. If (2b) is replaced by a dispersive wave packet of the form (2a) then the particles are trapped for a finite time only, and are eventually left behind by the wave (Knobloch and Weiss, 1987a).

The results described above are suggestive of the kinds of processes that are important for particle transport but are incomplete because the velocity field (2b) does not solve the nonlinear dynamical equations. Instead (2b) should be thought of only as the first term in an expansion of $u(x,t)$ in powers of A, *i.e.* it is valid only under the condition $|A| << c$. To discuss an example in which particle trapping is described **selfconsistently**, I turn to the recently discovered convective waves (Walden *et al*, 1985). In addition to providing a fully nonlinear analytical solution of the equations of motion, this example also demonstrates the importance of nonlinearities in selecting the wave *pattern*.

Convective waves arise in regions of a star that are stably stratified by a molecular weight gradient μ (due to ^4H_e, for example) and subject to a superadiabatic temperature gradient. In such a system an oscillatory diffusive instability, often called overstability, may occur even though the layer is statically stable. This process, known in astrophysics as semiconvection (Ulrich, 1972), occurs because the Lewis number $\tau \equiv \kappa_\mu/\kappa_{th}$ is less than unity (indeed, in the Sun, $\tau \sim 10^{-6}$). In this case a fluid element displaced upwards from its equilibrium position loses its heat faster than its H_e content, and consequently starts to descend. Since it is cooler when it returns to its original equilibrium position than on its way upward, it overshoots on its way down, and if $N_\mu > N_\mu{}^{crit}$, where N_μ is the Brunt-Vässäilä frequency due to the μ-gradient alone, the downward displacement will exceed the upward one. Thus growing oscillations will set in when the superadiabatic gradient, measured by the Rayleigh number R, exceeds a critical value R_0.

If the system is assumed to be translation and rotation invariant in the horizontal direction the general solution at R_0 to the linearized equations of motion is an arbitrary superposition of travelling waves :

$$\mathrm{w}_{\mathrm{lin}}\ (\mathbf{x}, z, t) = \Re_e \left(f(z) \int d\theta\ A(\theta)\ e^{i(\mathbf{k}.\mathbf{x}-\omega_0(\mathbf{k})t)} \right) \quad , \tag{7a}$$

where $\omega_0(\mathbf{k})$ is given by the dispersion relation, and the integral is carried out over all the $\mathbf{k}$ satisfying $|\mathbf{k}| = k_c$, the value minimizing R_0. Here w is the vertical velocity, $f(z)$ represents the common vertical structure of the eigenfunctions with $|\mathbf{k}| = k_c$ and θ denotes the orientation of k relative to some axis. Thus if $\mathbf{x} = (x,y)$, then $\mathbf{k} = k_c\ (\cos\theta,\ \sin\theta)$. The solution (7a) is neutrally stable. However, when R exceeds R_0 the solution will grow exponentially in time until nonlinear terms become important. When this is the case an arbitrary superposition of waves is no longer a solution and the solutions are greatly restricted in structure. When $0 < R - R_0 << R_0$ these nonlinear solutions may be approximated by particular solutions in the class (7a). Thus we may speak of the nonlinear terms as *selecting* a particular set of solutions from the far larger class of solutions represented by (7a). The significance of this observation becomes clear once it is realized that different waveforms mix and transport mass in very different ways.

To illustrate these points consider the two-dimensional problem in which a stably stratified fluid is confined between two unbounded horizontal planes and heated from below. Then (7a) is replaced by

$$w_{\text{lin}}\ (x,z,t) = \Re_e\ \left(A_1 e^{ik_c x} + A_2 e^{ik_c x}\right)\ f(z) \qquad , \tag{7b}$$

where

$$\begin{pmatrix} \dot{A}_1 \\ \dot{A}_2 \end{pmatrix} = \begin{pmatrix} i\omega_0 & 0 \\ 0 & -i\omega_0 \end{pmatrix} \begin{pmatrix} A_1 \\ A_2 \end{pmatrix} \qquad . \tag{7c}$$

Thus A_1 and A_2 represent left-travelling and right-travelling waves, respectively. For $R > R_0$ the spatially periodic solutions with the same wavelength $2\pi/k_c$ as the linear problem are determined by including in equation (7c) appropriate nonlinear terms. Then solving the equivalent of (7c) for A_1 and A_2 yields an approximation to the fully nonlinear problem in the form (7b), *i.e.*,

$$w_{\text{nonlin}}\ (x,z,t) = \ w_{\text{lin}}\ (x,z,t) + \ldots.. \qquad . \tag{8}$$

Recently developed theory (Knobloch et al, 1986; Knobloch, 1986a) shows that for $0 < R - R_0 << R_0$, equation (7c) can always be written in the form

$$\begin{aligned} \dot{A}_1 &= [\lambda + i\omega + a\ |A_2|^2 + b\ |A|^2 + \ldots]\ A_1 \qquad , & (9a) \\ \dot{A}_2 &= [\lambda - i\omega + \overline{a}\ |A_1|^2 + \overline{b}\ |A|^2 + \ldots]\ A_2 \qquad , & (9b) \end{aligned}$$

where $|A|^2 = |A_1|^2 + |A_2|^2$ is related to the Nusselt number, λ is proportional to $R - R_0$ and $\omega - \omega_0 = O(\lambda)$. The ellipses denote higher order terms in the amplitudes A_1, A_2. If we introduce real variables defined by $A_j = x_j\ \exp(i\theta_j)$, j=1,2, equations (9) become

$$\begin{aligned} \dot{x}_1 &= [\lambda + a_r\ {x_2}^2 + b_r\ ({x_1}^2 + {x_2}^2) + \ldots]\ x_1 & (10a) \\ \dot{x}_2 &= [\lambda + a_r\ {x_1}^2 + b_r\ ({x_1}^2 + {x_2}^2) + \ldots]\ x_2 \qquad , & (10b) \end{aligned}$$

for the real amplitude (x_1, x_2), with decoupled equations for θ_1, θ_2. Here the subscript r denotes the real part of the coefficients a, b.

The resulting equations have four types of solutions :
the trivial solution $(x_1, x_2) = (0,0)$ corresponding to pure conduction, $(x, 0)$ and $(0, x)$ corresponding to left- and right-travelling waves (**TW**), respectively, and (x, x) corresponding to standing waves (**SW**). These solutions are represented in the form of bifurcation diagrams $A(\lambda)$ in the (a_r, b_r) plane in fig. 1. Observe that stable solutions occur only if TW and SW both bifurcate supercritically (toward $\lambda > 0$) and that the stable solution (indicated by a solid line) is the one transporting more heat. Figure 1 points out quite clearly the role of the nonlinear terms in determining whether the instability evolves, for $R > R_0$, into a travelling or a standing wave. For example, in the region $\{a_r < 0, b_r < 0\}$ a solution in the form of SW will evolve into either a left- or right-travelling wave, depending on the nature of the perturbation. Since SW do not transport any mass, while TW are accompanied by both Lagrangian and Eulerian mean

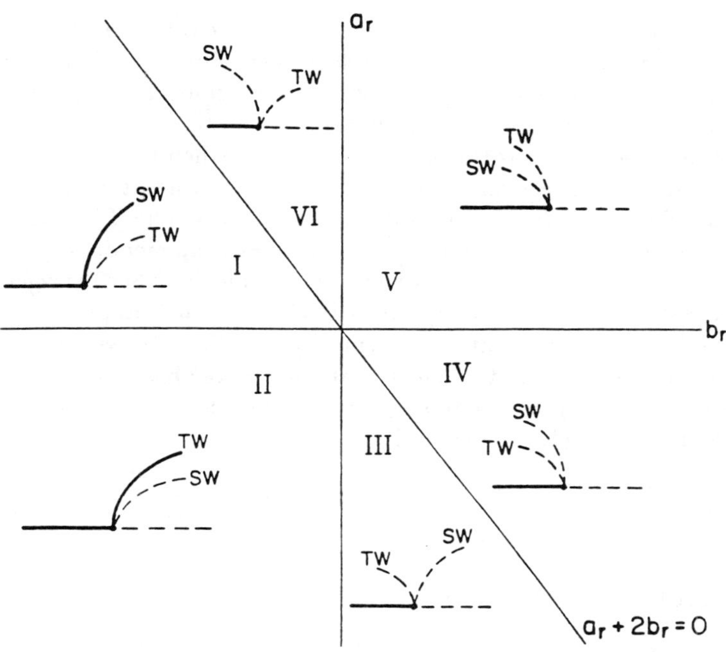

Figure 1. Bifurcations diagrams A(λ) from equations (10) in the (a_r, b_r)-plane. Stable (unstable) branches are indicated by solid (dashed) lines.

flows (cf. Knobloch and Moore, 1989), the theory presented above shows that it is the coefficients a_r, b_r that determine whether transport will take place or not.

The coefficients a_r, b_r depend on the boundary conditions that are used in conjunction with the partial differential equations describing the system. Without going into details, we list below the results of such computations. For thermosolutal convection (semi-convection) the preferred mode is a travelling wave (Knobloch et al, 1986); for rotating convection TW are preferred provided the Prandtl number $\sigma < 0.68$ and the Taylor number is sufficiently large (Knobloch and Silber, 1989); in contrast, for convection in a vertical magnetic field SW are preferred (Dangelmayr and Knobloch, 1986) although TW are preferred in a horizontal field (Knobloch, 1986b).

The above theory describes monochromatic wavetrains with unique spatial and temporal frequencies and their harmonics. These are the simplest nonlinear solutions of the governing partial differential equations. But more complicated nonlinear waves can also solve the governing equations. In particular we expect that as $R - R_0$ increases the pure wavetrain may undergo various secondary instabilities leading to more complex waveforms. Such solutions are in general hard to find, but become accessible to perturbation theory in special regions in parameter space, typically where the primary instability is degenerate. This is the case, for example, when the frequency ω_0 is close to zero. The analysis of the resulting "codimension-two" bifurcation shows (Dangelmayr and Knobloch, 1987) that for doubly diffusive convection the TW loses stability at a secondary oscillatory instability that introduces a new, independent frequency into the wavetrain without changing its spatial structure (Knobloch, 1986a). I refer to such waves as modulated travelling waves (**MW**). It is likely that such appearance of additional frequencies with increasing amplitude is typical of nonlinear waves, and for this reason I describe below in some detail the effect that such waves have on mass transport.

The bifurcation analysis provides a prediction for the form of the (scaled) streamfunction in a MW (Knobloch and Weiss, 1987b):

$$\psi(\xi,\zeta,t) = \psi_0(\xi,\zeta) + \delta\, \psi_1(\xi,\zeta,t), \quad |\delta| << 1 \quad , \tag{11a}$$

where

$$\psi_0(\xi,\zeta) = -\xi + R\cos\xi\sin\zeta \quad , \tag{11b}$$

$$\psi_1(\xi,\zeta,t) = \frac{1}{2}\left[(1-\frac{2}{\alpha})\cos(\xi+\alpha t) + (1+\frac{2}{\alpha})\cos(\xi-\alpha t)\right]\sin\zeta \quad , \tag{11c}$$

$\xi = x + ct$ is the comoving coordinate and $\zeta = \pi z$. When $\delta = 0$, (11) represents a pure left-travelling wave; when $\delta \neq 0$ the equation describes a modulated travelling wave. MW with $R = 3.266$ and $\alpha = 0.195$ have been observed in a recent experiment (Heinrichs et al, 1987).

Fig. 2a shows the instantaneous streamlines of ψ_0. When the TW amplitude $R > 1$ two stagnation points appear on both the top and the bottom boundaries; these are connected by a special streamline, $\psi_0 = 0$. Since this representation takes place in the comoving frame, all the fluid particles in the region enclosed by the streamline $\gamma : \{\psi_0 = 0\}$ travel more or less with the phase velocity c of the wave. These particles will be called "trapped". In contrast the particles outside γ drift backwards with respect to the comoving frame, and hence are called "untrapped". The appearance of the separatrix γ has important consequences for mixing. When $\delta \neq 0$, *i.e.*, when the wave is MW, the comoving streamfunction is no longer time-independent; consequently particles no longer follow the instantaneous streamlines. In fig. 2b I show the result of plotting the position of several fluid elements every modulation period $T = 2\pi/\alpha$ of the perturbation streamfunction ψ_1, obtained by numerically solving the equations

$$\dot{\xi} = -\partial\psi/\partial\zeta \ , \ \dot{\zeta} = \partial\psi/\partial\xi \quad . \tag{12}$$

This procedure defines the so-called time-T map. Observe that the trajectories corresponding to streamlines of ψ_0 that are far from the separatrix γ remain regular, while those near γ break up. Indeed it is possible to show (Knobloch and Weiss, 1987b) that when $\delta \neq 0$ the dynamics of (12) contains "horse-shoe" chaos. Near γ the fluid elements undergo a complex sequence of transitions between being trapped in the wave, and drifting backwards relative to it. The mixing resulting from this process has been studied in detail by Weiss and Knobloch (1989), and its complexity is illustrated in fig. 3 showing the interweaving of the region that is trapped for one iteration of the time-T map (fig. 3a) and the region which becomes untrapped after one iteration (fig. 3b). Using an approximate analytical form of this map calibrated against the solutions of (12), Weiss and Knobloch show that the process of repeated trapping and detrapping leads to **anomalous** diffusion, in the sense that

$$\Delta x^2 \equiv < (x(t) - x(0))^2 > - < x(t) - x(0) >^2 \sim t^{\nu}, \text{ as } t \to \infty \quad , \tag{13}$$

where, for the parameter values used, $\nu \approx 1.93$. Here the angled brackets indicate an ensemble average over the particles in the chaotic layer formed by the destruction of the separatrix γ. Properties of this type of chaos imply that asymptotically in time the particles in the layer are essentially perfectly mixed. It should be noted that owing to the long time tails in the distribution of trapped and untrapped particles it is not feasible to obtain an asymptotic result of the form (13) by direct integration of equations (12).

These results indicate that the phenomena responsible for mixing in this example are both complex and subtle, and require a careful study. In particular, in the present case a diffusion coefficient, as conventionally defined,

$$D = \lim_{t \to \infty} \Delta x^2 / t \quad , \tag{14}$$

does not even exist ! I suspect that this is a common feature of mixing by deterministic processes such as the one described here, and expect it to occur for more realistic forms of ψ, provided ψ_0 contains some type of separatrix.

From the point of view of mixing in the radiative core of the Sun, it is the internal gravity waves (g-modes) that are of greatest interest. It is important to apply some of the ideas and techniques developed for the convective waves described above to these waves. Press (1981) showed that the gravity waves excited in the convective overshoot will be focused towards the center of the Sun, where their amplitudes are sufficiently large that they may break (cf. Lindzen, 1981; McEwan, 1983), but did not address the mixing by such waves outside the nuclear-burning core. This process requires an understanding of the particle motions in such a wave. Since the convective overshoot excites internal wave packets rather than wave trains, and the wave packets propagate with the group velocity which is orthogonal to the phase velocity of the wave (Whitham, 1974) , the particle motions are considerably more complicated than those discussed above. In addition, it is important to appreciate the role of shear flows. The Sun does rotate differentially in radius, a fact that introduces critical layers into the flow. These occur when

$$u_0(z) = c \quad , \tag{15}$$

where $\mathbf{u} = (u_0(z), 0, 0)$ is the shear flow in (x, y, z) coordinates, and c the horizontal phase speed. Provided the Richardson number is of order one or greater, a wave propagating in such a flow will be effectively absorbed at its critical lever (Bretherton, 1966; Booker and Bretherton, 1967), and its energy and momentum converted, in the first instance, into horizontal streaming motion. When this gets too large the streaming is suppressed either by viscous dissipation (cf. Lin, 1967) or by nonlinear effects (Kelly and Maslowe, 1970). Thus in the presence of shear the wave energy and momentum

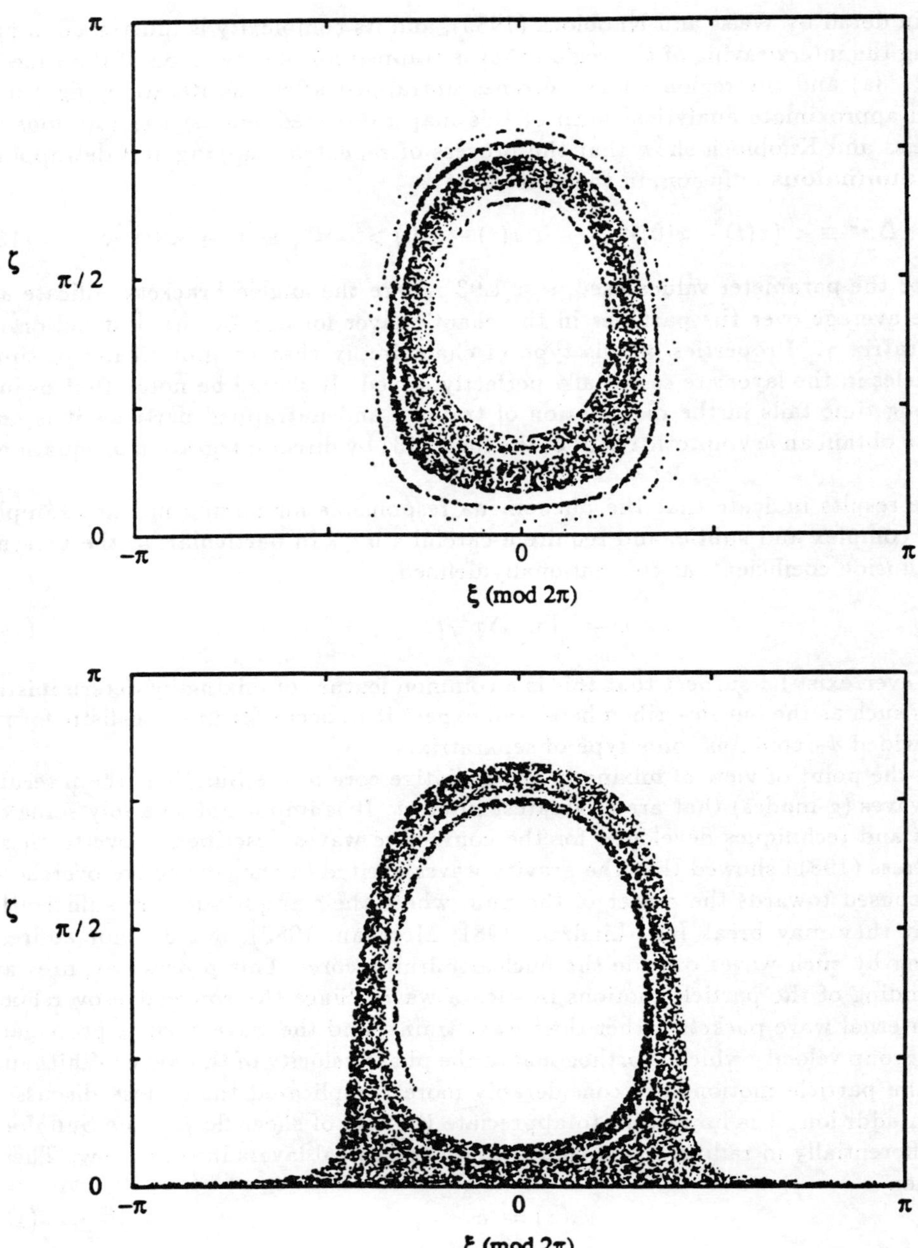

Figure 3. Regions which remain (a) trapped, and become untrapped (b) after one iteration of the time-T map, for parameter values of fig. 2(b).

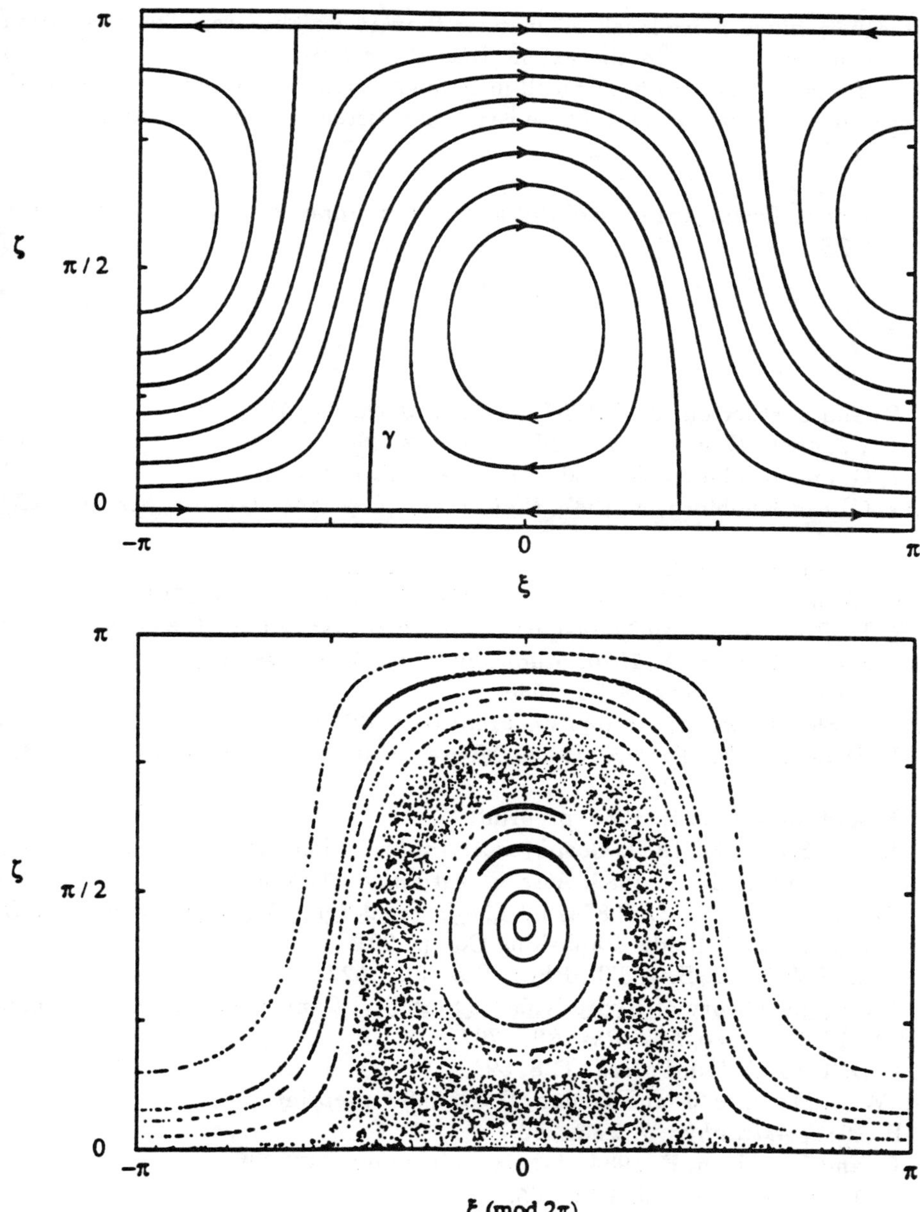

Figure 2. **(a)** The comoving streamlines in a pure travelling wave for R = 3.266, showing the separatrix γ separating regions of trapped and untrapped particles; **(b)** The time-T map for a modulated wave with R = 3.266, $\alpha = 0.195$ and $\delta = 0.4$ starting from several initial conditions.

are deposited in a much more localized manner than in its absence. The suggestion may therefore be made that, to the extent that the convective overshoot generates disturbances of a rather narrow bandwidth in frequency and wavenumber, local "jets" may be driven in the radiative core. These issues are currently under investigation.

Acknowledgement : The preparation of this article was supported in part by Cal Space, under grant # CS-11-89.

References

Booker, J. R. and Bretherton, F. P. 1967, *J. Fluid Mech.* **27**, 513.
Bretherton, F. P. 1966, *Quart. J. Roy. Met. Soc.* **92**, 466.
Dangelmayr, G. and Knobloch, E. 1986, *Phys. Lett. A* **117**, 394.
Dangelmayr, G. and Knobloch, E. 1987, *Phil. Trans. Roy. Soc. London*, Ser. A, **322**, 243.
Dilke, F. W. W. and Gough, D. O. 1972, *Nature* **240**, 262.
Heinrichs, R., Ahlers, G., and Cannell, D. S. 1987, *Phys. Rev. A* **35**, 2761.
Hurlburt, N. E., Toomre, J. and Massaguer, J. M. 1986, *Astrophys. J.* **311**, 563.
Kelly, R. E. and Maslowe, S. A. 1970, *Studies in Appl. Math.* **49**, 301.
Knobloch, E. 1986a, *Phys. Rev. A* **34**, 1538.
Knobloch, E. 1986b, *Geophys. Astrophys. Fluid Dyn.* **36**, 161.
Knobloch, E. Deane, A. E., Toomre, J. and Moore, D. R. 1986, *Contemp. Math.* **56**, 203.
Knobloch, E. and Moore, D. R. 1989, preprint.
Knobloch, E. and Silber, M. 1989, *Geophys. Astrophys. Fluid Dyn.*, in press.
Knobloch, E. and Spruit, H. C. 1983, *Astron. Astrophys.* **125**, 59.
Knobloch, E. and Weiss, J. B. 1987a, in The Internal Solar Angular Velocity (B. D. Durney and S. Sofia, eds), Reidel Publishing Co., p. 221.
Knobloch, E. and Weiss, J. B. 1987b, *Phys. Rev.* **36**, 1522.
Lin, C.-C. 1967, The Theory of Hydrodynamic Stability, Cambridge University Press.
Lindzen, R. S. 1981, *J. Geophys. Res.* **86**, 9707.
McEwan, A. D. 1983, *J. Fluid Mech.* **128**, 59.
Merryfield, W. J., Toomre, J. and Gough, D. O. 1989, preprint.
Press, W. H. 1981, *Astrophys. J.* **245**, 286.
Spruit, H. C. and Knobloch, E. 1984, *Astron. Astrophys.* **132**, 89.
Ulrich, R. K. 1972, *Astrophys. J.* **172**, 165.
Walden, R. W., Kolodner, P., Passner, A. and Surko, C. M. 1985, *Phys. Rev. Lett.* **55**, 496.
Weiss, J. B. and Knobloch, E. 1989, *Phys. Rev. A* **40**, 2579.
Whitham, G. B. 1974, Linear and Nonlinear Waves, Wiley.
Zahn, J. -P. 1987, in The Internal Solar Angular Velocity (B. D. Durney and S. Sofia, eds), Reidel Publishing Co., p. 201.

TURBULENT TRANSPORT IN STRATIFIED AND ROTATING FLUIDS

E. J. Hopfinger
Institut de Mécanique, UJF, INPG, CNRS, B.P. 53
38041 Grenoble

The surface abundances of Li, ^{13}C and other elements involved in the thermonuclear reaction in stars can only be exlained by hydrodynamic processes which lead to an effective transport of these elements well above molecular values. These mixing processes could be due to intermittent instability events as was suggested by Dilke and Gough (1972) or to more homogeneous, weakly turbulent motions. Schatzman (1977) outlined an approach for the determination of a turbulent diffusion coefficient in the stellar interior and he found that weak turbulence in stably stratified regions gives diffusivities that exceed the molecular values by about two orders of magnitudes. It might be entirely coincidental, but it is worth mentioning that vertical diffusivities in the stably stratified ocean interior are also, on average, about two orders of magnitudes above molecular values. Different types of instabilities occur in stars and lead to turbulence production. Convective instability is probably the most prominent one at least in certain regions. Doubly diffusive, baroclinic and shear instabilities are also possible . Schatzman (1977) in particular considered shear instability due to differential rotation as a likely candidate. In the sun, strong shears of this kind exist in the inner part (see Gavryuseva, 1986)

Localising and understanding instabilities, turbulence production and wave generation in stars are essential to the prediction of their evolution. In addition, it is necessary to know how turbulence is affected by stratification and rotation in particular when it is produced intermittently or when it penetrates into a stably stratified region. Three-dimensional turbulence under the effect of stratification or rotation degenerates into waves and quasi two-dimensional (2D) turbulence which interact. 2D turbulence is efficient in transporting mass in a direction perpedicular to gravity or the rotation axis. Transport parallal to gravity or the rotation vector would have to be accomplished by waves. This raises the important question about mass transport by waves discussed by Knobloch (this volume).

Effects of stratification on turbulence and mixing

The effect of stratification on turbulent transport is best illustrated by **figure 1**, origionally proposed by Posmentier (1977) and elaborated on by Linden(1979). For a passive scalar quantity (here temperature), of sufficiently weak gradient, the flux (expressed by Rf) is known to be proportional to the mean gradient (expressed by Ri), assuming of course that the turbulent flux can be represented by a gradient transport model. This implies that pertubations in mean gradient generated locally by the turbulent motions diffuse rapidly and the mean gradient tends to be smooth. When, for the same turbulent energy input the mean scalar gradient is increased such that it becomes dynamically active, the flux is no longer proportional to the gradient and for strong stratification is a decreasing function of the gradient. In this case local strong gradients will be enhanced and weak gradients will be further weakened. This process explains why the ocean microstructure consists of density steps adjacent to mixed regions.

More generally the heat or mass flux is expressed in relation to the turbulent kinetic energy production , namely by a flux Richardson number Rf, and the scalar gradient by a gradient Richardson number Ri as shown in **figure 1**. The maximum value the flux Richardson number can reach is about 0.3 at a gradient Richardson number of about 0.1 , meaning that at most 30% of the turbulent kinetic energy produced can be used for mixing; the reminder is dissipated by viscosity. The decrease in flux Richardson number when $Ri > 0.1$ is a consequence of the transfer of turbulent energy into internal wave motions which do little mixing or no mixing at all in the limit of linear waves.

Other than the gradient Richardson number which is expressed by external quantities such as the mean gradient and mean shear, local parameters characteristic of the turbulence structure, are also commonly used. These have a more universal character. When turbulence is created intermittently by a mean shear due to internal waves for instance which then ceases , as is often the case in stably stratified fluid systems such as the ocean, these intrinsic parameters are the only ones which allow to specify the state of evolution of the turbulence. The pertinent parameters are readily obtained from the transport equations for mean shear free turbulence (Hopfinger, 1987). By non-dimensinalising these equations by the length scale

$$L_\rho = -\overline{(\rho')^2}^{\,1/2}/(\partial\bar{\rho}/\partial z) \quad \text{and frequency} \quad N=(-(g/\bar{\rho})\,\partial\bar{\rho}/\partial z)^{1/2},$$

where ρ' is the density fluctuation, ρ the mean density and z is taken upward, opposite to the direction of gravity g, we obtain the following parameter:

$$N^3 w L_\rho / \varepsilon N \qquad (1)$$

which can be written in the form

$$(L_\rho / L_R)^2 (L_b / L_\rho). \qquad (2)$$

The scale $L_R = (\varepsilon/N^3)^{1/2}$ is known as the Ozmidov scale, $L_b = w/N$ is the buoancy scale, w the r.m.s.turbulent velocity parallel to gravity and ε the

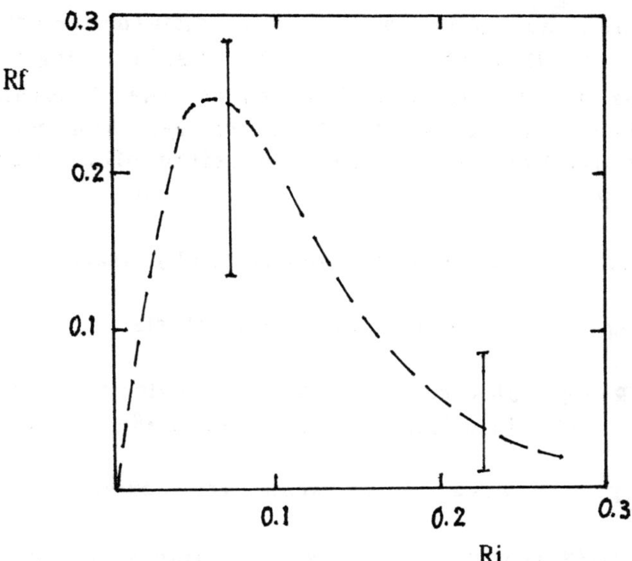

Figure 1 Turbulent heat or mass flux, expressed by a flux Richardson number Rf, as a function of scalar gradient, expressed by a gradient Richardson number Ri. Numerical values are taken from Linden(1979).

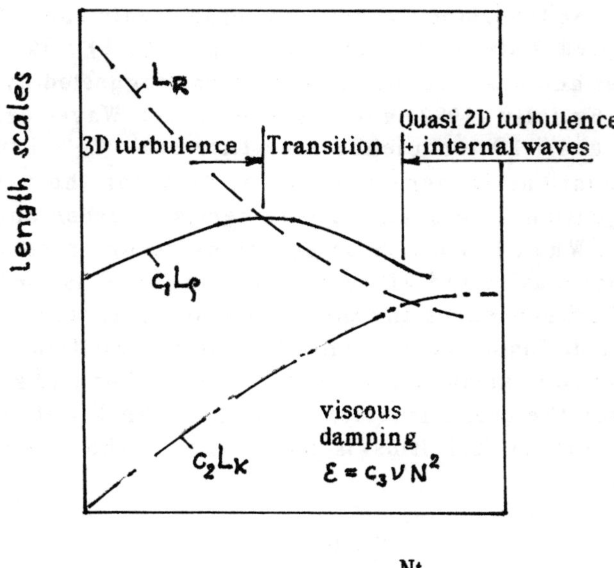

Figure 2. Schematic representation of the evolution of length scales L_ρ, L_R, L_K. The coefficients c_1, c_2, c_3 are determined experimentally.

dissipation rate. The ratio $L_b/L_\rho = w/NL_\rho$ is the disturbance Froude number F. From (1) it is seen that when $\varepsilon = w^3/L_\rho$, which is legitimate before turbulence collaps due to stratification begins, the disturbance Froude number is the only relevant parameter. Furthermore, when $w/NL_\rho = 1$ we have $L_R = L_\rho$. The effect of sratification is, therefore, often characterized in terms of the scale L_R :

when $L_R > L_\rho$, turbulence is unaffected by stratification,

$L_R \simeq L_\rho$, onset of stratification effects,

$L_\rho > L_R > L_K$, 3D turbulence exists on scales smaller than L_R and internal waves are generated at larger scales,

$L_R \sim L_K$, waves and 2D turbulence coexist.

The scale $L_K = (\nu^3/\varepsilon)^{1/4}$ is the Kolmogorov dissipation scale. The evolution of the scales as a function of time is sketched in **figure 2**. The coefficients c_1, c_2, c_3 in figure 2 are respectively of order 1.5, 10, and 15-30 (see Itsweire et al,1986). The turbulence length scale L_ρ is practically equal to the turbulent integral length scale L (Métais, 1985). The values of the non-dimensional times Nt which correspond to $L_R = c_1 L_\rho$ and to $L_R = c_2 L_K$ are Reynolds number dependent; in the laboratory values are respectively about 1.5 to 3 and 4 to 9.

The collapsed state of turbulence ($L_R \leq c_2 L_K$) is a superposition of internal waves and quasi 2D turbulence as was suggested by Riley et al (1981) and taken up by Lilly (1983) and Capéran (1989). Waves exist on a time scale $T_b = N^{-1}$ and quasi 2D turbulence on a time scale $T_a = L/u$. By writing the Boussinesq equations in non-dimensional form for the two time scales, two systems of equations are obtained with terms of order one and of order $F = u/LN$ and F^2 . When $F \to 0$, these equations reduce respectively to those of linear internal waves and 2D turbulence. When F is small but finite, the interactions between waves and turbulence occur at order F.

Evaluating diffusion coefficients in stably stratified fluids is a difficult task because of the variability of turbulence in time and space. Values quoted for example for the ocean interior range between $K = 10^{-4}$ to 10^{-7} m^2 s^{-1} .A way to determine diffusivities is by means of the Osborn and Cox (1972) expression

$$K = DC \quad , \tag{3}$$

where D is the molecular diffusivity and $C = \overline{(\partial \theta/ \partial z)^2} / \partial \overline{\theta}/ \partial z)^2$ is the Cox number. Gregg's (1980) measurements in the ocean microstructure give Cox numbers between 2 and 60, indicating that vertical diffusion of heat in the ocean interior is one to two orders of magnitude above molecular diffusion

This is of the same order as diffusivities in stars determined by Schatzman (1977).

Other expressions are of the form

$$K = k\,\varepsilon/N^2 \tag{4}$$

where k = 0.1 (Ozmidov, 1965). Expression (3) indicates that when vertical motions are absent we have K = D and expression (4) gives for vertical motions of order L_K a value of $K \sim \nu$, where ν is the kinematic viscosity. When the Prandtl number is very small, as is the case for stars, it is likely that (4) becomes meaningless. Concerning the usefulness of (3) it is difficult to see how Cox numbers could be obtained for stars.

Mixing across density interfaces

Often there exist well mixed regions, with nearly no mean gradient, adjacent to interfaces characterized by strong gradients. Examples of this are the ocean thermocline and the atmospheric inversion layer. Such interfaces seem also to exist in stars (J.-P. Zahn, private communication). A knowledge of transfer processes across interfaces is therefore also of interest in the astrophysical context. Let us consider the case without a mean shear where turbulent kinetic energy is produced at the base of the mixed layer at a rate q (**figure 3a**). Only a fraction of this energy will be available at the interface because most of it is dissipated by viscosity in the mixed layer(Hopfinger and Toly,1976). The important parameter is then the local Richardson number

$$Ri = g\Delta\rho L/\bar{\rho}\, u^2 \quad , \tag{5}$$

where u is the r.m.s. turbulent velocity at the interface and $\Delta\rho$ is the density jump across the interfaceof thickness h.

In laboratory experiments by E and Hopfinger(1986) turbulence was generated by an oscillating grid for which the turbulence structure, in particular u and the length scale L near the interface were known from previous measurements (Hopfinger and Toly). Erosion of the interface is accompanied by a slow increase in mixed layer depth H. This change in depth is at a time scale much less than any other time scale, say L/u, so that the process can be assumed to be quasi steady. The rate of change in H, that is, u_e = dH/dt, can be measured and is indicative of the mixing rate. Measurements of this quantity over a wide range of Ri led to the mixing law(E and Hopfinger)

$$u_e/u = a_1 Ri^{-3/2} \quad , \tag{6}$$

valid when Ri > 7. The upper limit of validity depends on Peclet number Pe = uL/D. The coefficient a_1 is of order 1. This mixing law is applicable when the

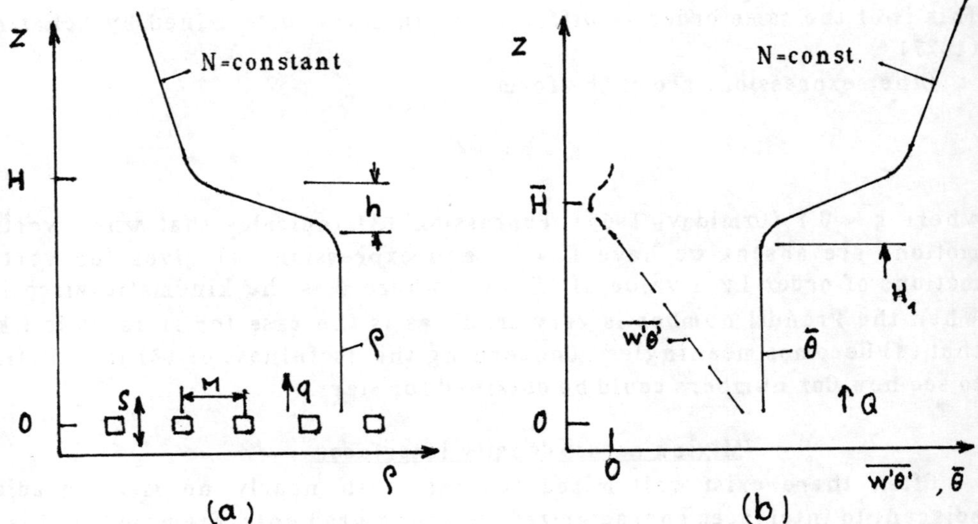

Figure 3. Configuration of mixing across interfaces; (a). mixing by oscillating grid produced turbulence; (b), penetrative convection.

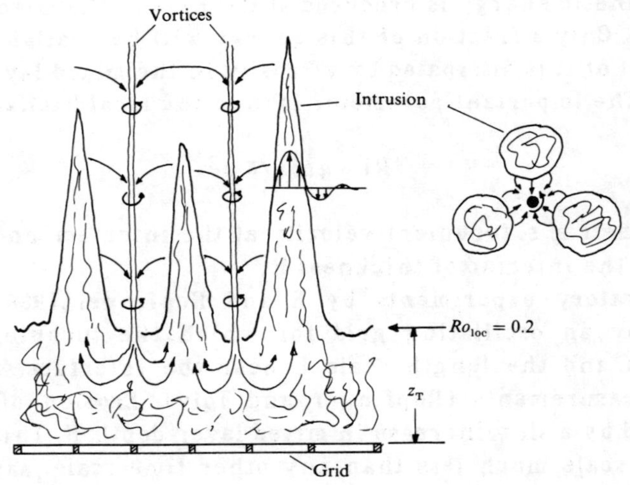

Figure 4. Formation of concentrated vortices by turbulence in rotating fluids. The rising clumns of mixed fluid cause convergence as sketched.

Peclet number is large (Crapper and Linden,1974; Piat and Hopfinger,1981), otherwise molecular diffusion must be taken into account. Hopfinger and Toly (1976) suggested the more general relation

$$u_e/u = Ri^{-3/2}(a_1 + a_2 (Ri/Pe)^{1/2}) \quad . \tag{7}$$

The $Ri^{-3/2}$ behaviour implies that the rate of mixing or rate of potential energy increase is not proportional to the rate of kinetic energy supply at the interface. Proportionality would lead to a Ri^{-1} behaviour. The $Ri^{-3/2}$ relation is obtained when it is argued that energy for mixing is made available at a buoancy or internal wave time scale $\tau \sim (\rho L/g\Delta\rho)^{1/2}$ rather than at the turbulent time scale L/u (Linden,1973). Other explanations have also been given(Mory,1989).

Penetrative convection follows similar laws. Deardorff, Willis and Stockton (1980) defined a Richerdson number $Ri^* = g\beta\Delta\theta H/w^{*2}$ with the turbulent velocity being given by $w^* = [g\beta\overline{(w'\theta')}_0\ H]^{1/3}$. The entrainment velocity was determined by $w_e = -\overline{(w'\theta')}_{\bar{H}}/\Delta\theta$. The indices 0 and H refer, respectively, to the origin z=0 and to the position of maximum negative heat flux at $z=\bar{H}$ (see **figure 3b**). β is the coefficient of thermal expansion and $\Delta\theta$ the temperature jump across the interface. Since $H \propto L$ and $w^* \propto u$ the Richarson number Ri^* is proportional to Ri defined in (5). The entrainment relation would, therefore, to be expected to be similar to (6). The experimental scatter of the data by Deardorff et al. is such that no clear entrainment law emerges. The data fall, however, within the bounds $w_e/w^* = a_3 Ri^{*-n}$ with $1 < n < 3/2$. The coefficient a_3 is again of order 1. Molecular effects seem to be negligible in the experiments by Deardorff et al., the Peclet number being $> 10^3$.

When the non-turbulent layer adjacent to the interface is stably stratified, as is often the case, internal waves can be excited in this layer which can radiate energy away from the interface and perhaps also transport mass. It has been shown by E and Hopfinger and also by Deardorff et al. that the energy radiated by internal waves is a very small fraction of the kinetic energy flux into the interface and the mixing rate is not altered by this energy leakage. Whether or not these waves can transport mass has not been investigated. It would be of interest to examine this question in the future.

The interface thickness is finite and E and Hopfinger established from experimental data the relation

$$h/H = 0.055 + 0.91\ Ri^{-1} \quad , \tag{8}$$

which is of the same form as Deardorff et al's. relation with the constants being somewhat different. This difference is in part due to a different definition of h and H. In any event, for large values of Ri the thickness is $\lesssim$

0.1H . The mechanism which determines this thickness is only speculative at present.

The above results are generally valid when there is no persistent mean shear. In stars, when the instability is due to differential rotation, a persistent mean shear is likely to exist. This changes the entrainment law and also the interface structure. It is threfore of interest to say a few words about this situation. When there is an interface some distance away from the shear plane, this interface acts on the mean flow in a way similar to a boundary. The mean flow under this constraint approaches the configuration of plane Couette flow (Piat and Hopfinger,1981) with strong mean shear at the boundaries and weak mean shear in the interior mixed layer. The density distribution is similar to mean-shear-free mixing.

The Richardson number in shear turbulence driven mixing is defined with the shear velocity u** and the mixed layer depth H in the form $Ri^{**} = g\Delta\rho H/\rho u^{**2}$. Momentum balance leads in this case to the relation(Piat and Hopfinger)

$$u_e/u^{**} = a_4 Ri^{**-1/2} \tag{9}$$

The coefficient a_4 is of order 1.

The interface thickness h adjusts to marginally stable conditions such that the gradient Richarson number $J = - g(\partial\bar{\rho}/\partial z)/\bar{\rho}(\partial U/\partial z)^2 < J_c$. Turbulence which diffuses from the mixed layer has a tendency to thin the interface provided $h/H \gtrsim 0.1$. When J falls below the critical value (about 0.3) because of the erosion by turbulence coming from outside the interface, the interfacial layer will become unstable and, as a consequence, thickening occurs. This competition between erosion and instability leads to a marginally stable state (Kantha et al.,1977, Piat and Hopfinger).

Rotation effects on turbulence and mixing

Rotation effects on turbulence and turbulent mixing become important when the turbulent Rossby number Ro = u/Lf (the analogue of F in stratified turbulence), where f is the Coriolis frequency 2Ω, is of order 1 or less. This was demonstrated by the experiments of Hopfinger et al.(1982). In these experiments oscillating grid turbulence of the type studied by Hopfinger and Toly(1976) was subjected to rotation with the rotation axis perpendicular to the grid plane. Without rotation the turbulent velocity u decreases with distance from the grid plane and the turbulent lengh scale increases linearely with distance. With rotation the turbulent Rossby number is large near the grid and decreases like z^{-2} away from the grid, where z is measured from the grid midplane. The experiments by Hopfinger et al. showed an abrupt change in turbulence structure at $z = z_T$, where Ro = 0.2 . It may be of interest to point out that, by analogy with stratified turbulence, the effect of rotation can also be expressed in terms of a lengh scale equivalent to the Ozmidov scale which in rotating fluid is (Mory and Hopfinger,1984)

$$L_{\Omega} = (\varepsilon/f^3)^{1/2} \qquad (10)$$

When Ro = 1, the scale $L_{\Omega} = L$ and when $L_K \sim L_{\Omega} < L$, or equivalently Ro < 0.2, the motion consists of quasi 2D turbulence and inertial waves, the anologue of internal waves in stratified fluid (Lighthill, 1978).

An interesting feature of rotating turbulence is the possibility of stong vorticity concentration into tornado-like vortices. These vortices support on them vortex waves (Hopfinger and Browand,1982) which themselfs transport energy. A possible mechanism of vorticity concentration is sketched in **figure 4**. Attempts to give theoretical explanations have been made by Mory and Capéran(1987) and by Goncharov and Gryanik(1986). Such vortices also form in high Rayleigh number thermal convection (Boubnov and Golitsyn,1986). Because of these vortices, rotatinally dominated turbulence is still capable of appreciable mass transport in a direction parallel to the rotation axis.

Experiments on mixing across a density interface by this rotationally dominated turbulence have recently been performed by Fleury et al.(1989). A density interface was introduced at $z < z_T$, where $0.2 < Ro < 1$, and at $z > z_T$, where $Ro < 0.2$. It was found that the mixing rate decreased with rotation according to the correlation

$$u_e/u = 0.5 \, Ro \, Ri^{-1} \qquad (11)$$

valid in the range $1 \ll Ri < 8 \, Ro^{-2}$. One explanation for the decrease in mixing rate is the energy radiation away from the interface by inertial waves as was suggested by Maxworthy(1986). Measurements of the kinetic energy in the non-stirred layer showed that the energy leakage can be large; the kinetic energy in the non-stirred layer can reach 30 to 40 % of the turbulent kinetic energy near the interface(Fleury et al.). This leakage alone can, however, not explain the observed decrease in mixing rate due to rotation. The change in turbulence structure and also the response of the interface to the forcing by the turbulent eddies are other possibilities. These aspects have recently been considered by Mory(1989).

References

Boubnov, B. M. and Golitsyn, G. S. 1986, J. Fluid Mech. **167**, 503.

Capéran, P. 1989, Thèse d'Etat, INPG.

Crapper, P. F. and Linden, P. F. 1974, J. Fluid Mech. **65**, 45.

Deardorff, J. W., Willis, G. E. and Stockton, B. H. 1980, J. Fluid Mech. **100**, 41.

Dilke, F. W. W. and Gough, D. O. 1972, Nature **240**, 262.

E, Xuequan and Hopfinger, E. J. 1986, J. Fluid Mech. **166**, 227.

Fleury, M., Mory, M. and Hopfinger, E.J. 1989,(Submitted to J. Fluid Mech.).

Gavryuseva, E. A. 1986, in <u>Proc. Workshop on Plasma Astrophys.</u>,ESA Publ.SP **251**,ISSN 0379-6566.

Gregg, M. C. 1980, J. Phys. Oceanogr. 10,915.

Goncharov, V. P. and Gryanik, V. M. 1986, Sov. Phys. JETP **64**, 976. ......

Hopfinger, E. J. and Toly, J. A. 1976, J. Fluid Mech. **78**, 155.

Hopfinger, E. J. and Browand,F. K. 1982, Nature **295**, 1.

Hopfinger, E. J., Browand, F. K. and Gagne,Y. 1982, J. Fluid Mech. **125**, 505.

Hopfinger, E.J. 1987,J. Geophys. Res. **92, NO. C5**,5287.

Itsweire, E. C., Helland, K. N. and VanAtta,C. W. 1986, J. Fluid Mech. **162**, 299.

Kantha, L. H., Phillips, O. M. and Azad, R. S. 1977, J. Fluid Mech. **79**, 753.

Knobloch, E. 1989 (this Volume).

Lighthill, J. 1978, <u>Waves in Fluids</u>, Cambridge University Press.

Lilly, D. K. 1983, J. Atmos. Sci. **40**, 749.

Linden. P. F. 1979,Geophys. Astrophys. Fluid Dyn. **13**, 3.

Métais, O. 1985,in <u>Proc. 5th Turbulent Shear Flows</u>, Cornell University, Ithaca,N.Y.

Mory, M. and Hopfinger, E. J. 1984, in Macroscopic Modelling of Turbulent Flows(Frisch, U. .... eds), Springer Verlag .

Mory, M. and Capéran, P. 1987, J. Fluid Mech. **185**, 121.

Mory, M. 1989 (submitted to J. Fluid Mech.).

Osborn, T. R. and Cox, C. S. 1972, Geophys. Astrophys. Fluid Dyn. **3**, 321.

Ozmidov, R. V. 1965, Izv. Acad. Sci., USSR, Atmos. Oceanic Phys. **1**, 493.

Piat, J.-F. and Hopfinger, E. J. 1981, J. Fluid Mech. **113**, 411.

Posmentier, E. S. 1977, J. Phys. Ocean. **7**, 298.

Riley, J. J., Metcalfe, R. W. and Weissmann, M. A. 1981, in <u>Proc. American Institute of Physics 76</u>, Nonlinear Properties of Internal Waves, New York, 79.

Schatzman, E. 1977, Astron. Astrophys. **56**, 211.

PENETRATION AND OVERSHOOTING FROM A CONVECTION ZONE

Josep M. Massager
Departament de Física Aplicada
Univ. Politècnica de Catalunya
08034 Barcelona, Spain

Abstract. Present knowledge on penetration and overshooting from stellar envelopes is reviewed and current assumptions are checked against laboratory experiments and numerical simulations. Emphasis is laid on quasi-adiabatic envelopes and, more precisely, on solar-type stars. Overshooting from the core is explicitly avoided.

1. Introduction

Proper determination of the mixed regions in stars is one of the major unsolved problems of stellar evolution theory. The most simple recipe, used in the early stages of stellar modeling, was to assume that for each stellar radius the system will choose either the adiabatic, ∂_A, or the radiative gradient, ∂_T, depending on which one of these takes the minimum absolute value

$$\partial_r T = \min\{-\partial_A,\ -\partial_T\}$$

where

$$\partial_T = -\,k^{-1}\frac{L}{4\pi r^2}$$

$$\partial_A = (1-\gamma^{-1})\ \frac{T}{P}\ \partial_r P$$

L, T and P are the luminosity, temperature and pressure, r is the stellar radius, $k = 4acT^3/3\chi\rho$ is the thermal diffusion coefficient and all other symbols retain their usual meaning. Assuming the regions where the temperature gradient is adiabatic to be well -instantaneously- mixed, the convective zone is identified with both the adiabatic and the mixed zones.

Proper understanding of how these three zones overlap is the major concern of the present paper. Local theories of convection, for instance, introduce a significant improvement by assuming for the temperature gradient, $\partial_r T$, a weighted average of ∂_A and ∂_T, with the averaging weights being functions of the local variables (Massaguer

1989, also §5). As a result, the convection zone can depart from adiabaticity as much as needed, and the edges of the convection zone, $\partial_T = \partial_A$, do not delimit the adiabatic zone. Yet still the adiabatic zone is included in between these edges.

Non-local theories address the question of how far beyond the edge of the convection zone mixing extends. This is certainly a precise question if, for instance, we agree in taking the edge at $\partial_T = \partial_A$. However, for practical purposes it can be a meaningless question, as will be shown below. If convection is strong enough to create an adiabatic region, as is often the case, a proper, and certainly much less precise question, should be asked. How far beyond the edge of the adiabatic zone -i.e.: the edge of the region where $\partial_r T \simeq \partial_A$ -does mixing extend?

The conceptual mismatch between the three zones -convective, adiabatic and mixing- is inherent in the use of local theories for modeling internal structure. Non-local theories participate in that confusion to some extent if they are conceived as small perturbations of a local model. Using the words penetration and overshooting as synonyms, as is usually done, is a consequence of this practice. The real breakthrough is to compute the whole mixed layer self-consistently, which is the main goal of some recently derived hydrodynamic models -see Massaguer's (1989) review for references. In the following we shall call overshooting a flow beyond the edge $\partial_T = \partial_A$ if it does not significantly change the mean temperature gradient -either because convection is weak or because the model is not self-consistent. Otherwise, we shall talk of penetrative convection.

The aim of the present paper is to lay the grounds of the subject by stressing comparisons between stellar observations, laboratory experiments and numerical simulations. None of these gives a completely satisfactory picture but all of them together give quite a reasonable description, and any newly derived hydrodynamic model should be tested against such a description. In this respect, the initiative taken by Tooth & Gough (1989) of checking their hydrodynamic model against laboratory experiments -even though they have only considered the non-penetrative case- is to be encouraged.

It will be shown below that penetration cannot be taken as an entity distinct from convection, nor can it be taken as a perturbation of convection between bounding walls. Penetration is the obvious outcome of convection in the absence of bounding walls. Therefore, convection in the core can hardly be treated as convection in the envelope. While the former flow is to be modeled as thermal convection in a fluid sphere with uniformly distributed heat sources, the latter is plain convection in a spherical shell. Also, the purpose of the present paper being to establish a connection as close as possible with physical reality, we shall avoid any further reference to overshooting or penetration from stellar cores. Because to the best of our knowledge no existing work, either laboratory or numerical, deals with penetrative convection in a sphere, with or without heat sources.

In the following we shall concentrate on the problem of penetration from stellar envelopes, which we believe to be very well described by a planar geometry. In §2 and §3 we shall review, respectively, what is known from laboratory and numerical experiments. Penetration from stellar envelopes will be discussed in §4. In §5 the most relevant ingredients for modeling penetrative convection will be summarized.

2. Penetrative convection as seen from experiments

Penetrative convection is a fairly frequent phenomenon in Nature. The rising of a nocturnal inversion layer in the Earth's atmosphere or the thermal convection under the thermocline in the upper ocean are examples of it. It has also been explored in some detail in laboratory experiments. Some of these experiments on penetrative convection give rise to a statistically steady flow, whereas others do not. The latter are the so-called run-down experiments because they evolve towards a final stage where convection mixes everything. They are barely indicative of the situation expected in stars but still give some hints.

2.1 *Run-down experiments*

The first experiment to be reviewed is that of Deardorff, Willis & Lilly (1969). They set up a constant temperature gradient in a tank filled with water. By increasing the bottom temperature at a constant rate the system developed a piecewise constant temperature gradient. The bottom layer was well mixed, displaying a constant temperature profile. On top of it the fluid layer kept its temperature profile unchanged, with the transition region between both profiles being smooth and very shallow. From this experiment it seems clear that, once the thickness of the whole layer is given, the knowledge of the top and bottom temperature plus that of the heat flux -i.e.: the temperature gradient in the conductive zone- is enough for the whole profile to be drawn. It must be noticed that if the Boussinesq fluid were to be changed into a compressible fluid, the constant temperature profile would to be changed into an adiabatic temperature gradient.

The paper quoted above does not give any information about the dynamics in the transition region. Whitehead & Chen (1970) conducted a very different experiment that gave a preliminary answer to this question. The experiment was not conceived as a run-down experiment, though it is not clear whether a statistically steady state could be achieved or not. The experiment consisted of heating the upper surface of a fluid layer by using a heating lamp. The upper layers of the fluid absorbed the radiated light, thus behaving as a fluid layer with heat sources. The conductive equilibrium obtained in such a way is unstable, with the unstable layer located on top of a stable one. The authors gave no indication about the mean structure that was finally obtained, but they reported on very intense dynamics in the interphase between the stable and unstable zones. Plumes, which they prefer to describe as jets, mixed the whole layer, and excited gravity waves in the lower stable region. Both phenomena, time-dependent plumes and gravity waves, will be shown to constitute the main signature of penetrative convection.

2.2 *Experiments on ice-water convection*

The only well-known experiment on penetrative convection that evolves towards a statistically steady state is that of the convection of water on top of an iced bottom. Conduction tends to create a linear temperature profile but, water displaying its highest density at $\bar{T} = 3.98^oC$, any cooler layer below that point is to be found unstable,

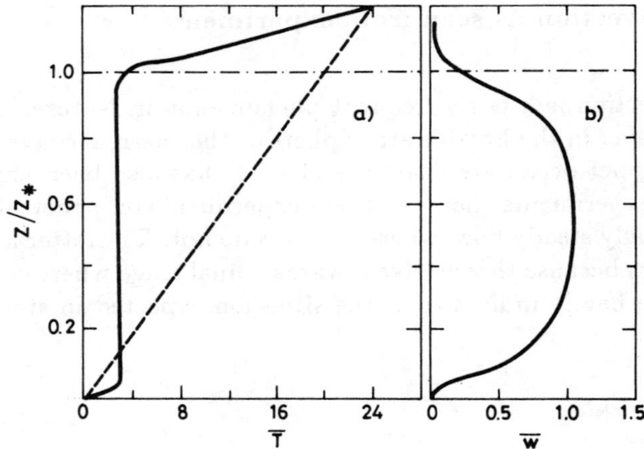

Figure 1. Laboratory measurements from an experiment on ice-water convection, re-drawn from Adrian (1975). The depth z/z_* is normalized to the point where the mean temperature takes the value $\bar{T} = 3.98^oC$. **a)** Mean temperature profile (solid line) and conductive temperature profile (dashed line) in degrees centigrade. The region of constant temperature gradient is the analog of the adiabatic zone in a stratified atmosphere. **b)** Best fit for the *r.m.s.* vertical velocity $\bar{w}$ in arbitrary units
(Courtesy of J. Fluid Mechanics).

so mixing tends to create a homogeneous temperature layer at, roughly, four degrees centigrade. Early experiments had been conducted by Townsend (1964) and Myrup *et al.* (1970). Both experiments showed intermittent narrow plumes extending across the mixed layer. In the former paper, temperature fluctuations of large magnitude near the interface of the unstable and stable layers were reported, and they were attributed to the breaking of internal gravity waves.

A later experiment on ice-water convection was conducted by Adrian (1975). It was mostly concerned with the detailed description of the interphase between the stable and unstable layers. Part of one of the figures in that paper has been re-drawn in figure 1. In figure 1(a) both a statistically steady mean temperature profile and the associated conductive temperature profile have been displayed. A roughly constant temperature layer can be easily identified. It is the analog of the adiabatic region for convection in a compressible fluid. Its extension is seven times larger than the originally unstable layer. This strength of the convective flow in flattening the entropy gradient comes in part from the fact that the ice-water convection displays a finite amplitude instability. As discussed by Moore & Weiss (1973), by conveniently distorting the temperature profile the flow can become unstable even at subcritical Rayleigh numbers. Thus, it does not come as a surprise that large distortions of the mean temperature profile arise even for mild supercritical Rayleigh numbers.

In Adrian's experiment, several momenta for the turbulent velocity and temperature fields were measured. His best fit for the mean vertical velocity field $\bar{w}$ against depth z/z_* has been plotted in figure 1(b), where, z_* is the depth of the temperature level

$\bar{T} = 3.98^oC$. Velocity decays in a very short distance and its amplitude is rather smaller in the transition region than in the bulk of the domain. In that paper, mixing in the region $z/z_* < 1$ is described as being done by coherent structures, which in turn excite internal gravity waves in the upper layer $z/z_* > 1$. The power-law estimated for the uppermost levels goes like $\bar{w} \; \alpha \; (z - z_*)^{-25/12}$, changing near the edge towards a $\bar{w} \; \alpha \; (z - z_*)^{-1/4}$ power-law.

3. Penetrative convection as seen from numerical simulations

Early attempts at modeling numerically penetrative convection involved ice-water convection. The first attempt was that of Musman (1968) by using the so-called mean-field approximation. Later on, Moore & Weiss (1973) produced a two-dimensional numerical simulation, and their results have been found consistent with Adrian's measurement. The only criticism may arise from their computed flow being cellular-like instead of plume-like, thus not giving the random time-dependence expected.

More recent attempts at modeling penetrative convection have been motivated by stellar structure theory. In the following we shall review a cooperative effort aimed at describing quasi-adiabatic convective envelopes, with the convective zone of the Sun as a target.

3.1 *Penetrative convection in a sandwiched layer*

The simplest way of setting up a penetrative convection flow is by putting three fluid layers one on top of the other, with the middle one being unstable and the other two adjoining layers being stable. In the numerical experiments to be reviewed below such a structure has been achieved by imposing either a piecewise constant conductivity -or opacity- or a piecewise constant adiabatic gradient. In the former case, and for a compressible fluid, the unperturbed sandwich displays a piecewise polytropic structure. Both issues have been examined and found to be consistent.

A major difficulty with the assumed model comes from taking either the conductivity or the adiabatic gradient -i.e.: the magnitude defining the change in stability- to be a function of the geometrical depth. This is certainly an unphysical assumption. Proper modeling should assume either magnitude to be a function of temperature and density, but this would possibly imply unwanted additional instabilities such as the kappa-mechanism. As a consequence, in these papers the edge of the convection zone, $\partial_T = \partial_A$, was pinned at a fixed position, thus enhancing the relevance of the edge of the unstable layer. As an additional, though minor consequence, the finite amplitude instability described by Moore & Weiss (1973) for ice-water convection is inhibited, although other physical mechanisms producing finite amplitude instability, like the stratification itself (Massaguer & Zahn 1980), can still be active.

3.1.1 *On modeling cellular convection*

The first attempts at modeling penetration in a sandwiched layer were based on a modal expansion, and most of the results were obtained for a single horizontal mode representation, thus imposing on the flow a well-defined cellular structure. In the following we shall only consider three-dimensional cells -i.e.: hexagonal planforms- as they give the most plausible results. In such a geometry the streamlines become more concentrated near the centreline of the cell than near its border, thus showing a preferred orientation. This is not a crucial issue in Boussinesq convection, as the equations there display a mirror symmetry against a mid-plane, and every orientation is equally alike but it certainly matters in a stratified fluid. In the following we shall call the flow upwards or downwards according to orientation of the velocity field along the centreline.

The most important question for astrophysical applications concerns the thickness of the adiabatic zone. Roughly speaking, for a Boussinesq fluid the entropy gradient changes discontinuously from its radiative value to a zero gradient value (Zahn, Toomre & Latour 1982). The adiabatic zone is certainly larger than the unstable layer, so penetration is important, but the temperature gradient across the whole sandwich is either radiative or adiabatic with, roughly speaking, no intermediate values.

In a stratified fluid and for a down mode -i.e.: centrewards- the phenomenology is very similar (Massaguer *et al.* 1984). Yet now, if the density contrast between top and bottom is very large, the upper section of the unstable layer can depart significantly from the adiabatic gradient. These layers are under the influence of an additional stabilization effect induced by the so-called buoyancy inversion (Massaguer & Zahn 1980), with the flow being squeezed downwards, as can be seen in figure 2(*b*.) It can also be realized from this figure how sharp the edges of the adiabatic zone are, thus giving this zone a well-defined entity.

The upward modes display a very different phenomenology, as their penetration is almost negligible, with the edges of both the convective and the adiabatic zones being superimposed. But what does come as a real surprise is that once both up and down modes act together, though interacting only through the mean structure, the upwards mode dominates, thus giving negligible penetration. Such a lack of penetration can be questioned on physical grounds, as it works against any laboratory-minded intuition. However, as both types of modes give two different types of energetic balance, proper sorting of the dominant mechanism matters. Two-dimensional numerical simulations to be reviewed below give the opposite answer, thus favouring the down-mode balance, and therefore displaying important penetration.

Besides the temperature gradient, the velocity field is the most important magnitude to be measured in a penetrative setting. The thickness of the non-adiabatic mixed layer has been found to be very shallow in every computed model, with the velocity there being a small fraction of its maximum value in the bulk of the domain, as can be seen from figure 2. Such a result seems to be consistent with laboratory measurements -see figure 1(*b*). Yet the large values displayed by the gradient of the horizontal velocity, see figure 2, can be suspected of being unstable if the truncated model could properly deal with shear instabilities. In fact, the flow displayed in figure 1 is time-dependent and plume-like, while that in figure 2 is steady and cellular. Thus we can expect this non-adiabatic mixed regions to display shear-induced turbulence.

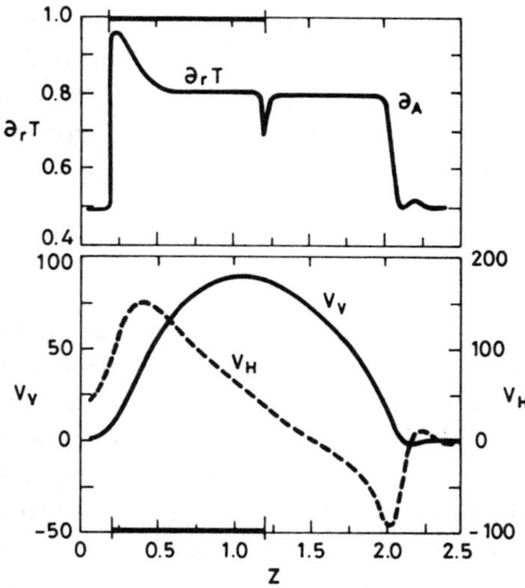

Figure 2. Results from a numerical simulation of convection in a highly stratified fluid on the assumption of anelastic modal expansion. Re-drawn from Massaguer *et al.* (1984) the case of a downwards hexagon mode. The thick line on the axis of abscissae defines the unstable layer. **a)** Mean temperature gradient $\partial_r T$ as a function of depth. The value of the adiabatic temperature gradient is $\partial_A = 0.8$. The spike at the edge of the unstable zone is forced by the discontinuity imposed on the thermal difusion coefficient. **b)** Vertical velocity V_V (solid line) and horizontal velocity V_H (dashed line). *(Courtesy of Astron. & Astrophys.)*

3.1.2 *Modeling penetration from a two-dimensional approximation*

The highly truncated models described above raise some question about the dominant balance and, also, about the structure of the mixed layer. Some preliminary answers can be obtained from the two-dimensional numerical simulation carried out by Hurlburt, Toomre & Massaguer (1986), even if a two-dimensional approximation is far from being not free from criticism.

What is new and physically satisfactory in this work is its plume-like, time-dependent flow. A major consequence of the non-cellular geometry of the flow is an intense time-dependent dynamics, with plumes entraining ambient fluid to their interior. As the ambient fluid is at rest, the entrainment results in a damping, which slows down the motion. Therefore, as compared with a cellular down flow, penetration decreases significantly. The main balance is no longer between buoyancy work and kinetic energy production, and entrainment of ambient fluid by shear also plays an important role.

An additional point to be mentioned is the presence, in the lower stable layer of internal gravity waves excited by the plumes themselves. It is very difficult to differentiate the region where the velocity field produces mixing from that where the velocity field is made up only of waves, if indeed this difference could be made. Every statistical momentum -i.e.: mean kinetic energy, mechanical flux, etc.- computed in this region decreases in a different length-scale. A preliminary attempt at properly defining a mixed region can be found in Hurlburt, Toomre & Massaguer (1989), where the dynamics of a passive scalar in a two-dimensional flow is examined. But proper understanding of the internal gravity waves' transport properties is a non-trivial, as yet unsolved problem, which deserves profound theoretical effort (Knobloch 1989).

3.1.3 *Upwards penetration*

Most of what has been said above concerned downwards -i.e.: centrewards - penetration from a convective envelope. Penetration towards the upper layers of a highly stratified atmosphere, say towards the photosphere in the solar case, is quite different. There, the main balance is different from that in a laboratory setting. Non-Boussinesq effects can be very important. The stabilizing effect of the pressure fluctuations, induced by buoyancy inversion and shear, constrains the motion to be cellular instead of plume-like. Penetration, measured in geometrical length is very small, and the edges of the convection and adiabatic zones are located at the same place. Penetration is almost reduced to an undulation of the top boundary. However, if the adiabatic zone is taken as the physically relevant entity, it can be realized that mixed regions on top and below the adiabatic zone span a similar geometrical depth. What makes a real difference between both mixed-layers, however, is their strength in producing internal gravity waves, which is much smaller in the upper layer than in the lower one.

3.2 *Convection in an A-type star*

As regards convection, stellar envelopes range from those where advection is highly efficient in transporting heat, thus making the temperature gradient adiabatic almost everywhere, to those where heat transport by advection is negligible. Roughly speaking, the effective temperature plays the role of an "order parameter" in Thermodynamics. In main sequence stars, decreasing the effective temperature increases the efficiency of convection. In the hottest stars displaying a convective zone the mean temperature gradient is independent of convection, and is so for the whole structure of the star. But the hydrodynamic activity can still be very large there.

The prototype of an inefficient convective envelope is that of an A-type star (Toomre, Zahn, Latour & Spiegel 1976). These stars show, in fact, two such convective shells, with penetration tunneling between them both (Latour, Toomre & Zahn 1981). Their structure is almost insensitive to the details of convection except, perhaps, as regards those processes associated with changing the chemical composition by mixing. The description that will be given below follows from the two papers quoted above. They are based on a modal expansion of the anelastic equations. Results were obtained for both, two-dimensional rolls and three-dimensional hexagons, but only the latter will be considered here.

As anticipated, the entropy profile for the computed solutions did not show any significant departure from the profile obtained on the assumption of radiative equilibrium. Convection was so mild that it did not flatten the mean entropy profile

$\bar{S}(r)$. Thus, these stars provide a typical example of overshooting. The extent of penetration, as measured from the edge $\partial_T = \partial_A$,or $\partial_r \bar{S} = 0$, to the first zero of the vertical velocity field was found to be, roughly speaking, one scale-height, which in turn is equivalent to the whole depth of the unstable layer. In spite of the reduced thermal activity, as measured in terms of the convective heat transport, the mixed layer was found to be the site of an important hydrodynamic activity. It is noteworthy that the region separating the two unstable layers could be tunneled by convection, thus completely mixing the overlapping region.

Although the above description is in complete agreement with current knowledge on A-type stars, comparison with the laboratory and numerical experiments quoted above raises some doubts. The existence of a convective zone with an almost conductive gradient is, certainly, completely consistent with the experiments on Benard convection, but it might not be with experiments on penetrative convection. If the temperature contrast across a fluid layer in radiative equilibrium is increased monotonously, the radiative equilibrium will become unstable and a convective flow will be triggered. In Rayleigh-Benard convection the amplitude of the velocity at the onset of instability is negligible and increases monotonously with the temperature contrast, thus reaching first a small amplitude velocity regime. But penetrative convection arises as a finite amplitude instability, as discussed by Moore & Weiss (1973) for ice-water convection -see also §2.2. Therefore, it is not clear why an almost radiative regime is to be reached if penetration is allowed. To be precise, penetrative convection arises as a finite-amplitude bifurcation, and there exist two branches of possible solutions, which can both be reached in a numerical experiment. The structure and stability of such branches may depend on details, and the existence of a new branch of quasi-adiabatic solutions cannot be ruled out.

4. Penetrative convection from quasi-adiabatic envelopes

The prototype of a star with a quasi-adiabatic envelope is the Sun. But in spite of the unquestionable efforts that have been devoted to such a purpose, no attempt at producing a hydrodynamic model has succeeded as yet, mostly because of the strong numerical constraints imposed by its high density stratification, but also because of parameterization difficulties.

Observations, however, have made it possible to do what numerical simulations have failed to do. Recent work by Christensen-Dalsgaard, Gough & Thomson (1989) has shown a well-defined adiabatic zone located in the top twenty-three percent of the solar radius. They have obtained the temperature distribution from the speed of sound, measured inside the Sun by inverting helioseismological data. Thus, their results have to be taken as a direct measurement of the temperature gradient. By plotting the curve $W = \frac{r^2}{Gm} \partial_r c^2 \alpha \partial_r \bar{S}$ against the solar radius they found a flat region with $W \simeq 0$ on the upper layers of the Sun, but the steepness of the curve changed discontinuously at $r/R_\odot = 0.77$, with W increasing with depth. Therefore the edge of the adiabatic region can be identified very easily, but there was nothing that could be identified as the edge $\partial_T = \partial_A$. We must conclude either that this edge coincides with the end of the adiabatic zone $\partial_r T \simeq \partial_A$ or that its position is of no practical relevance, as emphasized in previous discussions of laboratory experiments and numerical simulations.

From the point of view of stellar modeling, the question concerns the calibration of local models and the relationship of such a calibration with the depth of the adiabatic zone. Mixing-length theories, for instance, are dependent on the so-called mixing-length parameter, which measures the ratio of the mixing length to the pressure scale-height. Calibration means choosing the adiabat so as to fit some constraints -say the radius of the star for a fixed chemical composition. But choosing the adiabat also means choosing the depth of the adiabatic zone. Therefore, it is inconsistent with local modeling to assume for the adiabatic zone an extension beyond the edge $\partial_T = \partial_A$. The convective flow can, indeed, overshoot that edge but certainly not modify the entropy profile below that point.

5. Ingredients for a theory

In the previous sections it has been shown how every experiment on penetrative convection, either laboratory or numerical, gives a very similar picture, with the sole exception of the model of an A-type star reviewed. This picture consists of a layer of fluid displaying a quasi-adiabatic temperature gradient and bounded by two adjoining layers in radiative equilibrium. The mixed layer slightly overflows the adiabatic zone, thus giving two very shallow overshooting layers. If the density contrast between top and bottom is very large, then the uppermost levels in the adiabatic zone behave as a transition layer, with the gradient there being intermediate between its radiative and adiabatic values.

In spite of the apparent simplicity of the model, there still remains one major problem to be solved. How can the adiabat for the convective zone be properly chosen? Local models do so by calibration (Gough & Weiss 1976) or by giving an asymptotic power law for the Nusselt number N (Massaguer 1989), which comes to be the same thing. To be precise, the temperature gradient can be written as the weighted average

$$\partial_r T = \frac{N-1}{N}\,\partial_A + \frac{1}{N}\,\partial_T$$

where N is to be written as a function of the Rayleigh and the Prandtl numbers. If most of the heat flux is transported by convection then $N \gg 1$, implying $\partial_r T \simeq \partial_A$. In the opposite case, $N \simeq 1$, then $\partial_r T \simeq \partial_T$. Therefore, a precise law for convection is required only for the transition layer. But this is the region where non-local and non-Boussinesq effects can play a dominant role. Therefore, the answer is not an easy one.

5.1 *On non-local modeling*

There are several useful short-cuts in order to derive a model that can be plugged into a stellar evolutionary code. For instance, to derive a hydrodynamic model from the full equations by imposing some *ad hoc* restrictions, often as closures on low order momenta, or by also deriving some asymptotic relationships. The rationale for such approximations originates in the equations themselves but, in order to be effective, the approximations are usually so crude that a large amount of phenomenology is required.

A different short-cut is taken by non-local modeling, as its equations only express the main balances. Mass conservation is imposed by assuming a lagrangian description -i.e.: plumes, thermals, eddies etc. However, as mentioned in §3.1, entrainment of material into the lagrangian element is to be included. It reduces the specific momentum content and, by this bias, the balance between buoyancy work and kinetic energy production in a parcel of fluid is broken.

Proper modeling of the advection of material into a fluid element is difficult, but it can be eased by using the experience obtained from modeling plumes in laboratory experiments (see, for instance, Turner 1973). A first attempt at modeling plumes in a compressible fluid or in a stratified atmosphere can be found in Schmitt, Rosner & Bohn (1984).

Non-local models are often used also to estimate the properties of the velocity field below the adiabatic zone. In §2, results from experiments in ice-water convection have been quoted. They at least give some hints about scale-lengths for velocity to decay. But, as discussed in connection with the numerical experiments described above, decaying lengths are different for each one of the velocity momenta. As the plumes swinging induces random-like velocity fields, even order momenta, such as the kinetic energy, decay in a much longer distance than odd order momenta, such as the mechanical flux. Therefore, only soundly established turbulence models are suitable in order to describe the overshooting from a quasi-adiabatic region.

6. Conclusion

In the present paper the structure of a convective zone in a penetrative setting has been examined. Laboratory experiments, numerical simulations and observational results have been compared one against the others in order to draft a general picture. A consistent scenario has been produced for stars showing a well-developed convection zone, which has been called a quasi-adiabatic convection zone. Convection in stars showing inefficient convective zones has also been discussed, but results concerning this case are too scarce to be cross-checked and some questions still remain open.

The present work was supported by the Dirección General de Investigación Científica y Técnica (DGICYT) under grant PS87-0107.

References

Adrian, R. J. 1975, *J. Fluid Mech.* **69**, 753.

Christensen-Dalsgaard, J. Gough, D. O. and Thompson, M. 1989, in Inside the Sun, G. Berthomieu and M. Cribier Eds., Kluwer.

Deardorff, J. W., Willis, G. E. and Lilly, D. K. 1969, *J. Fluid Mech.* **35**, 7.
Gough, D. O. and Weiss, N. O. 1976, *Mon. Not. Roy. Astron. Soc.* **176**, 589.
Hurlburt, N., Toomre, J., and Massaguer, J. M. 1986, *Astrophys. J.* **311**, 563.
Hurlburt, N., Toomre, J., and Massaguer, J. M. 1989, "Penetration and Mixing Below a Convection Zone" (preprint).
Knobloch, E. 1989, (this meeting).
Latour, J., Toomre, J. and Zahn, J. P. 1981, *Astrophys. J.* **248**, 1081.
Massaguer, J. M. 1989, in Inside the Sun, G. Berthomieu and M. Cribier Eds., Kluwer
Massaguer, J. M., Latour, J., Toomre, J. and Zahn, J.P. 1984, *Astron. Astrophys.* **140**, 1.
Massaguer, J. M. and Zahn J. P. 1980, *Astron. Astrophys* **87**, 315.
Moore, D. R. and Weiss, N. O. 1973, *J. Fluid Mech.* **61**, 553.
Musman, S. 1968, *J. Fluid Mech.* **31**, 342.
Myrup, L., Gross D., Hoo, L. S. and Goddard W. 1970, *Weather* **25**, 150.
Schmitt, J. H. M. M., Rosner, R. and Bohn, H.U. 1984, *Astrophys. J.* **282**, 316.
Toomre, J., Zahn, J. P., Latour, J. and Spiegel, E. A. 1976, *Astrophys. J.* **207**, 545.
Tooth, P. D. and Gough, D. O. 1989, in Inside the Sun, G. Berthomieu and M. Cribier Eds., Kluwer.
Townsend, A. A. 1964, *Quart. J. Roy. Met. Soc.* **90**, 248.
Turner, J. S. 1973, Buoyancy effects in Fluids, Cambridge.
Whitehead, J. A. and Chen, M. M. 1970, *J. Fluid Mech.* **40**, 549.
Zahn, J. P., Toomre, J. and Latour, J. 1982, *Geophys. & Astrophys. Fluid Dynamics* **22**, 159.

TURBULENT TRANSPORT IN STELLAR RADIATION ZONES: CAUSES AND EFFECTS

Jean-Paul Zahn
Observatoire Midi-Pyrénées
31400 Toulouse, France
and
Columbia University
New York, NY 10027, USA

Abstract. We briefly review the instabilities which are the most likely to generate turbulent transport of chemicals and of angular momentum in radiative stellar interiors, namely the shear instabilities due to differential rotation. Estimates are given for the turbulent diffusivity, and it is examined how a meridional circulation can cause such differential rotation.

1. Introduction

Evry Schatzman was the first to point out that the chemical composition at the surface of the Sun and solar-like stars can only be interpreted by a mild but efficient transport of matter, which must take place well below their convective envelope, in the stable radiative region (Schatzman 1969, 1977; Schatzman and Maeder 1981).

Such transport can of course be achieved by large-scale circulations, for instance through the well known Eddington-Sweet circulation which is generated by the thermal imbalance of a rotating star (Eddington 1925, Vogt 1925, Sweet 1950, Mestel 1953). Another large-scale motion, the so-called Ekman circulation (Ekman 1905), probably occurs at the bottom of the solar convection zone, as suggested already by Bretherton and Spiegel (1968).

But Evry Schatzman had a different explanation. His intuition was that the transport of chemicals he was considering is due to motions of smaller scale, which he assumed to be caused by hydrodynamical instabilities. He characterized this weak turbulence by a Reynolds number Re^*, and proceeded to deduce from the observations the turbulent diffusivity $D_t = \nu Re^*$ (ν being the kinematic viscosity). The Re^* he found for the depletion of lithium et beryllium at the solar surface were rather small, below the critical Reynolds numbers required to trigger such instabilities.

In spite of this apparent paradox, his intuition proved to be right. In fact, both large-scale circulations and turbulent mixing are likely to operate in the radiative interior; they compete and they are probably related. Such circulations advect angular momentum and therefore they induce a non-uniform rotation which is highly supercritical. The kinetic energy which is fed into that differential rotation can thus be released by various hydrodynamical instabilities, and be transformed into turbulent motions whose vertical component may well be subcritical. It is the purpose of the present paper to discuss this chain of processes.

Before we do this, let us recall that the radiative interior of the Sun is in almost uniform rotation, as we recently learned through helioseismology (Brown *et al.* 1989); this indicates that there must be also some transport of angular momentum, in addition to that of matter. Those two transports need not to be linked, but it is worth examining whether they cannot be ascribed to the same cause.

2. Properties of shear instabilities

The radiative interior of a star is stable with respect to the convective instability and it is therefore considered, to first approximation, as being in a state of no motion (except for the rotation). It is well known, however, that a whole set of instabilities can occur there, due to various causes. Among them, it appears that the so-called shear instabilities are the most likely, if not the only ones, to reach a state which may be described as turbulent, and that they are able to actually mix the stellar material (see Knobloch and Spruit 1982, 1983; Zahn 1983).

The simplest fluid motion which displays such instabilities is a plane-parallel flow whose constant downstream velocity U varies across the stream, in the z-direction. The strength of the shear is measured by the gradient dU/dz, which is also the magnitude of the vorticity.

In the absence of other restoring forces, it is observed that all such flows are unstable, once the Reynolds number which characterizes them becomes large enough, although the detailed properties of the instability depend on the profile $U(z)$. When U has an inflexion point somewhere in the domain, corresponding to a maximum of vorticity, it has been shown by Rayleigh (1880) (see also Fjørtoft 1950) that the flow is unstable to infinitesimal perturbations. When the flow does not display such an inflexion point, it generally requires a perturbation of finite amplitude to become unstable. The theoretical study of that non-linear case is more intricate; it is found that the instability criterion does not only imply the strength of the perturbation, but also its shape (Gill 1965, Lerner and Knobloch 1988, Dubrulle and Zahn 1990).

But shear flows can be stabilized by a restoring force, and the most common example of this is provided by the buoyancy. In a stably stratified medium, that restoring force is characterized by the buoyancy frequency N, which in a homogeneous star is given by

$$N^2 = \frac{g}{H_P}(\nabla_{ad} - \nabla_{rad}), \qquad (1)$$

with the usual notations for the gravity, the presssure scale height, and the logarithmic temperature gradients of stellar structure. If the perturbations associated with the motions were adiabatic, the shear instability would be prevented as soon as

$$\frac{N^2}{(dU/dz)^2} > 1/4\,, \tag{2}$$

a condition known as the Richardson criterion.

In a star, however, the buoyancy force is diminished through radiative damping, because momentum diffuses much slower than heat; the ratio σ between those two diffusivities, which is called the Prandtl number, is very small in stellar interiors. It can be shown that the stability criterion above then takes the form (Zahn 1974)

$$\frac{N^2}{(dU/dz)^2}\,\sigma R_c > 1\,, \tag{3}$$

with R_c designating the critical Reynolds number for that shear profile.

However, when the star is not homogeneous but the molecular weight increases with depth (so that the stratification remains stable), the buoyancy force is only partly reduced through radiative damping; then the original Richardson condition (eq. 2) again applies, with a modified buoyancy frequency

$$(N_\mu)^2 = \frac{g}{H_P}\,\frac{d\ln\mu}{d\ln P}\,. \tag{4}$$

Such a gradient of molecular weight can be achieved through nuclear reactions, or through gravitational settling.

A similar restoring force operates in a rotating fluid, namely the Coriolis force, but it does not play the same role as the buoyancy. This was shown for an inviscid fluid by Johnson (1963), who found that the most unstable modes are not affected by the rotation. Experiments have been performed with real fluids, for instance by Rabaud and Couder (1983), and they clearly confirm that the onset of the shear instability is insensitive to the Coriolis force.

Those properties of plane-parallel shear flows can be readily applied to differentially rotating stars. Since a star is stratified, we must consider separately the vertical differential rotation, where the angular velocity Ω varies with depth r, and horizontal differential rotation, where Ω varies with latitude. In the first case, the buoyancy acts as a stabilizing force, and the instability criterion which holds for a homogeneous star is (from eq. 3)

$$\left(s\frac{\partial\Omega}{\partial r}\right)^2 > N^2\,\sigma\,R_c \tag{5}$$

(s being the distance to the rotation axis). When there is a gradient of molecular weight, one recovers the Richardson criterion with the modified buoyancy frequency

$$\left(s\frac{\partial\Omega}{\partial r}\right)^2 > 4\,(N_\mu)^2\,. \tag{6}$$

In the second case, that of horizontal differential rotation, the shear instability cannot be prevented by the stratification, since the buoyancy force acts only in the vertical direction. We may thus conclude that, due to the very large Reynolds number characterizing these flows in stars, a differential rotation in latitude is always unstable (Zahn 1975).

When such a horizontal shear flow becomes unstable, it first gives rise to motions which have the same vorticity as the mean flow, and which are therefore horizontal, and two-dimensional. Those motions may be turbulent eddies, or they may be waves (inertial waves in this case, since the Coriolis force is the major force acting in the horizontal direction). Those horizontal motions may, in turn, undergo a three-dimensional instability, unless the vertical motions are prevented by some restoring force. In any case, the resulting motion field is highly anisotropic, and one has to take this property into account when estimating the transport efficiency of those motions, as will be done next.

For completeness, we must mention the possibility of yet another restoring force, which is due to the magnetic field. This Laplace-Lorentz force will probably play a major role near the bottom of a convection zone, which is thought to be the site of the dynamo mechanism (see Spiegel 1987). But if the field alternates its polarity in time, as in the Sun, it will not penetrate deep into the star, and it should not interfer much there with the differential rotation and with the shear instabilities mentioned above. The possibility of a fossile field traversing the star is not very likely either; it is hard to see how it would survive the pre–main-sequence phase and the various instabilities which tend to destroy it (see Tayler 1982).

3. Their effect: turbulent transport

Let us now examine how a turbulent velocity field achieves the transport of a scalar quantity such as temperature, the concentration of chemical species, etc. (The definition of the term "turbulent" is out of the scope of this paper - let us just take it here in its intuitive sense.) To first approximation, this transport can be described as a diffusion process (see Knobloch 1978), provided some conditions are met (a sufficient scale separation, for instance). The situation is more complex when dealing with vector fields, because those are also subjected to stretching, which may lead to enhancement of the fields, not only their diffusion. When the velocity field is isotropic, the turbulent diffusivity reduces to a scalar, D_t, which is given by

$$D_t = \frac{1}{3} u\ell , \tag{7}$$

where u is the r.m.s. velocity and ℓ the correlation length of the turbulent velocity field. In the general case, when the turbulent motions are anisotropic, D_t is a tensor. In the phenomenogical approaches, these quantities are viewed as the velocity and the size (or the mixing length) of the turbulent eddies.

To estimate the mixing length ℓ, several prescriptions are available. Sometimes one can just take the dimension of the turbulent region, when the largest eddies are actually of that size. But in many instances the kinetic energy is injected at a scale which is smaller than that of the whole unstable region; the most vigorous eddies, which contribute most to the turbulent transport, are then of intermediate size. It is for that reason that, when modelling stellar convection zones, one relates the mixing length to the local pressure scale height, following E. Vitense (1953). In the same spirit, one can choose for ℓ the distance to the nearest boundary, as was done originally by Prandtl.

Likewise, several prescriptions are avalaible to estimate the velocity u. The largest turbulent velocities are often comparable with the variation ΔU of the mean flow speed over the unstable domain. In many cases, however, that amplitude ΔU is not prescribed, but results from the turbulent regime, whose strength in turn is determined by another requirement, such as the flux of momentum or of heat that has to be transported. When it is possible to directly estimate the rate ε_t at which kinetic energy is injected into the turbulent motions at the scale ℓ, it can be readily established that

$$u \approx (\varepsilon_t \, \ell)^{1/3} \tag{8}$$

(see Landau and Lifschitz 1987). An alternate approach is to estimate u through the growth rate $1/\tau$ of the considered instability :

$$u \approx \frac{\ell}{\tau} , \tag{9}$$

that growth rate being derived from the linear instability theory. The correct procedure is to calculate its value once the instability has reached its full amplitude and has modified the mean fields. For instance, in the classical mixing-length treatment for convection the velocity is estimated by following the acceleration (over a mixing length) of an eddy in the actual superadiabatic stratification that has been established by the convection; the expression for the convective flux is then found to be

$$F_c \approx \rho u^3 \frac{H_P}{\ell} . \tag{10}$$

Note that this expression yields the kinetic energy injection rate u^3/ℓ in terms of the convective flux.

Let us now consider specifically the turbulence which is generated in the horizontal differential rotation. We already mentioned that the motions resulting from that shear instability are strongly anisotropic: they are more vigorous and of larger extent in the horizontal direction than in the vertical. The diffusion coefficient is thus no longer a scalar, but a tensor.

To determine the turbulent diffusivity in the vertical direction, we proceed as follows. We have seen that the horizontal shear gives rise to horizontal motions, which themselves may be unstable and yield three-dimensional eddies. But this transition from 2 to 3 dimensions can be hindered by a restoring force.

Among such forces, the Coriolis force plays a major role: it dominates the dynamics of the largest motions, a property which is well known in geophysical fluids, and which it is illustrated by various laboratory experiments (see Hopfinger *et al.* 1982). Those large-scale motions will remain horizontal and two-dimensional, and they may also take the form of inertial waves; in any case, they will not contribute to the turbulent transport in the vertical direction. However, those large-scale motions have a very large Reynolds number and therefore they cannot dissipate the kinetic energy which is fed into them by the shear instability. We are thus led to the conjecture that they must couple with small-scale, turbulent motions which are liable to viscous friction. If some conditions are met, those smaller eddies have a turn-over rate which is larger than the rotation frequency, and therefore they are not dominated by the Coriolis force; we may thus assume that they are three-dimensional, and that they obey the Kolmogorov law (see Landau and Lifshitz 1987). When it exists, this three-dimensional tail of the turbulence starts at the scale which verifies both and

$$u'/\ell' \approx \Omega \quad \text{and} \quad (u')^3/\ell' \approx \varepsilon_t ; \tag{11}$$

it follows that the vertical turbulent diffusivity is given by

$$D_v \approx u'\ell' \approx \varepsilon_t/\Omega^2 . \tag{12}$$

Of course, when $D_v < \nu$, there are no three-dimensional motions, and the vertical diffusion reduces to the microscopic one (which is of the order of the kinematic viscosity ν).

The buoyancy is the other restoring force which is always present in a star, but it operates here only in a vertical gradient of molecular weight, since the temperature perturbations associated with those turbulent motions are completely smoothed out by radiative damping. Such a μ-gradient will inhibit three-dimensional turbulence for the eddies whose turn-over rate is less than the residual buoyancy frequency (eq. 4)

$$u/\ell < N_\mu , \tag{13}$$

and it will suppress it entirely when (see Zahn 1983)

$$(N_\mu)^2 > \varepsilon_t/\nu . \tag{14}$$

A rather small gradient of molecular weight is thus sufficient to prevent turbulent diffusion in the vertical direction, such as that due to the varying composition of ^{3}He in the Sun.

So far we have focused our attention on the vertical transport which can be accomplished through turbulent motions. Another possibility exists, namely transport by waves, as discussed by E. Knobloch in this volume. Such waves are probably emitted by the shear instabilities, as has been already mentioned, but they are also due to other mechanisms (coupling with convective motions, for instance). One interesting property of transport by waves is that it can ensure the redistribution of angular momentum in a star without affecting its chemical composition. Indeed, waves may traverse regions in which there is no turbulent mixing; for instance, isothermal gravity waves propagate in a region of stable μ-stratification which, as we have just seen, does not allow for turbulent diffusion (Zahn 1989). This could explain why the particle transport occurs on a much smaller rate than the momentum transport, as was pointed out by Poinsonneault *et al.* (1988). This whole subject of transport by waves is still in infancy, but it appears extremely promising.

4. One possible cause: meridional circulation

Let us come back to turbulent diffusion, whose efficiency, as we have seen, mainly depends on the injection rate ε_t of the turbulent energy. An obvious source of such energy in a radiative stellar interior is the kinetic energy of differential rotation, which can be easily released through the shear instabilities that have been discussed. That differential rotation can be due to several causes: the contraction or expansion of the star during its evolution; angular momentum loss through a wind; the presence of a differentially rotating convection zone, which induces an Ekman circulation; thermally driven meridional circulation; even tidal braking in a binary star. The case we shall treat here is that of differential rotation induced by a meridional circulation $\mathbf{U}$.

Such a circulation advects angular momentum, according to

$$\frac{\partial}{\partial t}\left(s^2\Omega\right) + \mathbf{U}\cdot\nabla\left(s^2\Omega\right) = \Gamma, \tag{15}$$

where Γ is the torque exerted per unit mass by the turbulent motions (as before, s is the distance to the rotation axis). Likewise, we may describe the evolution of the rotational kinetic energy:

$$\frac{\partial}{\partial t}\frac{1}{2}\left(s\Omega\right)^2 + \Omega\,\mathbf{U}\cdot\nabla\left(s^2\Omega\right) = \Omega\,\Gamma. \tag{16}$$

In a stationary state, the advection of kinetic energy is balanced by the work done by the turbulent torque, and this is precisely the quantity that we wish to determine, namely the injection rate of kinetic energy into the turbulence. Splitting the angular velocity into its mean component (over a level surface) and the horizontal differential rotation $\delta\Omega(r,\theta)$ (r being the mean radius and θ the colatitude), we obtain the following expression for the turbulent energy input, averaged over a the whole horizontal layer (Zahn 1987):

$$\varepsilon_t(r) = -\int_0^1 \delta\Omega(r,\theta)\,\mathbf{U}\cdot\nabla\left(\Omega\sin^2\theta\right)\,d(\cos\theta). \tag{17}$$

To proceed further, one needs a prescription to determine the strength of the horizontal differential rotation $\delta\Omega$. This is presently the weakest point of the theory, for we do not know what actually governs the distribution of angular momentum over a level surface (turbulent diffusion or transport by waves?). In the meanwhile, all we can do is to assert $\delta\Omega = -C\,\Omega_0\,P_2(\cos\theta)$, introducing a parameter C which we may calibrate by comparing with the observations. Replacing the circulation velocity U by its actual value, it is then possible to give an estimate of the energy generation rate ε_t and of the vertical diffusivity D_t.

For the Eddington-Sweet circulation, those quantities may be crudely approximated by

$$\varepsilon_t \approx C\frac{L}{M}\left(\frac{\Omega^2R^3}{GM}\right)^2 \quad \text{and} \quad D_v = \frac{\varepsilon_t}{\Omega^2} \approx C < K > \frac{\Omega^2R^3}{GM}, \tag{18}$$

where $< K >= LR^3/GM^2$ is the mean thermal conductivity of the star.

Turbulent viscosity thus operates on a timescale which is of the same order as that of the Eddington-Sweet circulation, which implies that the large scale advection will be affected by that turbulent diffusion. (That value of D_v given above must be considered as a rough estimate; in fact, it varies with depth and depends strongly on the rotation law.)

Note that the turbulent diffusivity scales as the square of the rotational velocity; if it is indeed responsible for the transport of chemicals, the observations should reflect that dependence on the rotation of the star. When taking the coefficient C of order unity, D_v appears to have the magnitude of what is required to interpret the depletion of Li in solar-like stars (Baglin *et al.* 1985, Vauclair 1988); it may also explain the abundance of Li observed in the old halo stars (Vauclair 1988).

The turbulent transport associated with the Ekman circulation below the convection zone has not been thoroughly examined so far. The velocity of that circulation depends on the turbulent viscosity within the boundary layer in which the rotation adjusts itself to that imposed by the convective stresses. It is thus crucial to determine whether this circulation penetrates into the convection zone, as was assumed by Bretherton and Spiegel (1968), where it would experience a very high diffusivity. If this is not the case, the turbulence is due to the instability caused by the vertical shear, and the diffusivity takes a much smaller value. Another question to settle is that of the vertical extent of the circulation, a question which has been also examined by Spiegel (1987). Until those uncertainties are cleared, it will be difficult to assess the efficiency of the Ekman circulation in the radiative interior.

5. Conclusion

As anticipated by Evry Schatzman, turbulent diffusion appears to play a major role in mixing the chemical species within the radiative interior of a star; we propose that it is caused by the instabilities due to the ever present differential rotation. This transport process carries to the surface some elements which are produced in the deep interior, such as ^{3}He in the Sun (Geiss *et al.* 1972), or ^{13}C in the giant stars of the first ascending branch (Lambert *et al.* 1980). On the other hand, it moves fragile elements from the surface to a depth where they are destroyed, and it is probably responsible for the depletion of Li observed in solar-like stars (Boesgard 1976, Duncan 1981, Cayrel *et al.* 1984). For the same reason, it may also affect the abundance of Li in the old halo stars, and therefore such observations (Spite and Spite 1982) must be treated with caution when taken as a test for the validation of cosmological theories. Turbulent diffusion competes also with other processes (transport by waves, magnetic torquing) in distributing angular momentum within a star, and it may be suppressed by a sufficient gradient of molecular weight.

The theory of that turbulence is still in a very crude state. The weakest point is our poor knowledge, at present, of the horizontal transport of angular momentum, which governs the rate at which kinetic energy is transferred from the differential rotation of the star into the turbulent motions. One way to progress, and to guide the theoretical work, is to implement the few recipes which have been derived into evolutionary models, and to compare the theoretical predictions with the observations. This route was taken by Schatzman and Maeder (1981), Baglin *et al.* (1985), and more recently by Vauclair (1988), Pinsonneault *et al.* (1989), and by Charbonneau and Michaud (1990). Some discrepancies which have been noticed are certainly due to physical processes that have not been taken in account so far, such as the momentum transport by waves, or the effect of an Ekman circulation at the bottom of the convection zone.

Evry Schatzman truly desserves our gratitude for opening such an exciting new field in the stellar structure theory!

Acknowledgements. I wish to thank all the participants of our meeting in Les Houches for the interest they expressed in this work, and in particular E. Hopfinger, E. Knobloch, E. Schatzman, H. Spruit and S. Vauclair, with whom I discussed many aspects of the problem. Partial support by the C.N.R.S. and by the U.S. Air Force (grant AFOSR 89-0012) is kindly acknowledged.

References

Baglin, A., Morel, P. and Schatzman, E. 1985, *Astron. Astrophys.* **149**, 309.

Boesgard, A. 1976, *Publ. Astron. Soc. Pacific* **88**, 353.

Bretherton, F.P. and Spiegel, E.A. 1968, *Astrophys. J.* **153**, 277.

Brown, T.M., Christensen-Dalsgaard, J., Dziembowski, W.A., Goode, P., Gough, D.O. and Morrow, C.A. 1989, *Astrophys. J.* **343**, 526.

Charbonneau, P. and Michaud, G. 1990, *Astrophys. J.*, **352**, 681.

Cayrel, R., Cayrel de Strobel, G., Campbell, B. and Däppen, W. 1984, *Astron. Astrophys.* **283**, 205.

Dubrulle, B. and Zahn, J.-P. 1990, submitted to *J. Fluid Mech.*.

Duncan, D.K. 1981, *Astrophys. J.* **248**, 651.

Ekman, V.W. 1905, *Arkiv Matem. Astron. Fysik* **2**, 11.

Eddington, A.S. 1925, *Observatory* **48**, 78.

Fjørtoft, R. 1950, *Geofys. Publ.* **17**, 52.

Geiss J., Buhler, F., Cerutti, H., Eberhardt, P. and Filleux, C.H. 1972, Apollo 16 Prel. Sci. Rep. NASA (SP 315).

Gill, A.E. 1965, *J. Fluid Mech.* **21**, 503.

Hopfinger, E.J., Browand, F.K. and Gagne, Y. 1982, *J. Fluid Mech.* **125**, 505.

Johnson, J.A. 1963, *J. Fluid Mech.* **17**, 337.

Knobloch, E. 1978, *Astrophys. J.* **225**, 1050.

Knobloch, E. and Spruit, H.C. 1982, *Astron. Astrophys.* **113**, 261.

Knobloch, E. and Spruit, H.C. 1983, *Astron. Astrophys.* **125**, 59.

Lambert, D.L., Dominy, J.F. and Sivertsen, S. 1980, *Astron. Astrophys.* **235**, 114.

Landau, L. and Lifschitz, E. 1987, Fluid Mechanics (English translation, 2nd edition; Pergamon).

Lerner, J. and Knobloch, E. 1988, *J. Fluid Mech.* **189**, 117.

Mestel, L. 1953, *Mon. Not. Roy. Astron. Soc.* **113**, 716.

Pinsonneault, M.H., Kawaler, S.D., Sofia, S. and Demarque, P. 1989, *Astrophys. J.* **338**, 424.

Rabaud, M. and Couder, Y. 1983, *J. Fluid Mech.* **136**, 291.

Rayleigh, Lord 1880, *Scientific Papers*, **1**, 474 (Cambridge Univ. Press).

Schatzman, E. 1969, *Astrophys. Lett.* **3**, 139.

Schatzman, E. 1977, *Astron. Astrophys.* **56**, 211.

Schatzman, E. and A. Maeder 1981, *Astron. Astrophys.* **96**, 1.

Spiegel, E.A. 1987, The Internal Solar Angular Velocity (eds. B.R. Durney and S. Sofia; Reidel), 321.

Spite, F. and Spite, M. 1982, *Astron. Astrophys.* **115**, 357.

Sweet, P.A. 1950, *Mon. Not. Roy. Astron. Soc.* **110**, 548.

Tayler, R. J. 1982, *Mon. Not. Roy. Astron. Soc.* **198**, 811.

Vauclair, S. 1988, *Astron. Astrophys.* **335**, 971.

Vitense, E. 1953, *Z. Astrophys.* **32**, 135.

Vogt, H. 1925, *Astron. Nachr.* **223**, 229.

Zahn, J.-P. 1974, Stellar Instability and Evolution (eds. P. Ledoux, A. Noels and R.W. Rodgers; Reidel), 185.

Zahn, J.-P. 1975, *Mém. Soc. Roy. Sci. Liège, 6e série* **8**, 31.

Zahn, J.-P. 1983, Astrophys. Processes in Upper Main Sequence Stars (Publ. Observatoire Genève), 253.

Zahn, J.-P. 1987, The Internal Solar Angular Velocity (eds. B.R. Durney and S. Sofia; Reidel), 201.

Zahn, J.-P. 1989, Inside the Sun (eds. G. Berthomieu and M. Cribier; Kluwer), 425.

WILL A MAGNETIC FIELD INHIBIT TURBULENT TRANSPORT?

H.C. Spruit
Max Planck Institut für Physik und Astrophysik
Karl-Schwarzschildstr. 1, D-8046 Garching

Introduction

The observed Lithium depletion of the Sun indicates the presence of some form of mixing below the base of the convection zone. The depth of the convection zone, now well measured by helioseismology (Christensen Dalsgaard *et al.* 1990) is 200.4 Mm. This is about 25 Mm above the depth where Lithium burns. The weak mixing required to carry the Lithium in the convection zone down to this depth is plausibly attributed to hydrodynamic processes associated with differential rotation (Schatzman, 1977). Measurements of the internal rotation of the sun (Brown *et al.* 1989) show that the latitude dependent rotation in the convection zone changes into a more uniform rotation below. The Lithium burning layer lies in the transition region. The difference between the polar and the equatorial rotation rate in this region may produce two dimensional turbulence (on equipotential surfaces). The tail of the turbulent spectrum at small scales would be three dimensional, and cause mixing, as described by Zahn in this volume.

At the same time however, the measured internal rotation rate has a very weak radial gradient. To maintain such a small gradient in the presence of the solar wind torque acting continously on the solar surface (Schatzman, 1962) a very effective angular momentum transport mechanism must be present in the core. If this mechanism is parametrized by an effective viscosity, its value must have been about 100 times larger than the turbulent diffusivity indicated by the Lithium depletion (Law *et al.* 1984). The mechanism, if it is responsible for both the Lithium depletion and the low rotation gradient, is apparently very *anisotropic*. In a stably stratified medium such as the solar core, anisotropic transport in the sense required is possible, as evidence in the earth oceans shows (e.g. Tassoul and Tassoul, 1989). The anisotropy is intimately related in this case with the presence of internal gravity waves (see Knobloch, this volume; Hopfinger, this volume).

In addition to such purely hydrodynamic mechanisms for angular momentum transport, a very natural mechanism is the torque exerted by a weak magnetic field. Only a field of the order of 1 Gauss is required. Such a weak field is for all practical purposes unobservable by means other than their effect on the internal rotation, so that one must rely on theoretical arguments about its likely strength and configuration. In the following, I review the relatively simple arguments that have been made so far. They indicate that a field strength of 1G or higher is quite plausible. Such a field may start interfering with hydrodynamic processes, so that the question arises whether the mixing below the convection zone must be treated as a hydrodynamic or as an essentially *magnetohydrodynamic* problem.

Visibility at the surface of a field in the core

A convectively driven dynamo process maintains a field in the lower part of the convection zone. Somewhat indirect plausiblity arguments give it a strength of the order 10^4 G, in a layer with thickness of the order 10 Mm near the base (the product of these quantities being constrained by observations at the surface). If a field unrelated to this dynamo process were present in the core and if it were to penetrate the convection zone, it would cause an asymmetry between consecutive activity cycles because the polarity of the dynamo generated field changes between cycles. Such an asymmetry is known to exist. Unfortunately, there are also processes internal to the activity cycle that can cause such an asymmetry, such as the ubiquitous period doubling occuring in dynamic systems, so that this observation can not be used. Theoretical considerations of the interaction of magnetic fields with convection favor the idea that an external magnetic field would be expelled by the convection zone (Parker, 1963; Drobyshevskii and Yuferev, 1974; Galloway and Weiss, 1981; Vainshtein *et al.* 1972). The field in the core could then be up to 10^4 G (the value at which its energy density becomes comparable with the kinetic energy density in the convection zone) without showing up at the surface.

By analogy with the many known magnetic A stars (sharing many properties with the solar core) the core might have a strong field, up to several kG. A significant difference between the A stars and the sun however is that the A stars have a convective core which the sun does not. Since it is still possible that the field of the magnetic A-stars is generated by a dynamo process in their convective cores, this makes the A star model of unreliable applicability.

Effect of the field on hydrodynamic processes

The magnetic field starts interfering with a hydrodynamic process when the magnetic energy density becomes comparable with the kinetic energy of the process in the absence of a field. This is equivalent to the Alfvén speed being of the same order as the flow speed. Since the magnetic effects are very anisotropic however, it pays to be a bit more precise about this. Consider two kinds of flow. In the first, the flow in a plane perpendicular to the field lines is purely compressive (curl free), in the second purely rotational (divergence free). A general flow can be decomposed into these two kinds. In both cases we construct the flow parallel to the field such that the divergence of the full three dimensional flow is zero, because we have applications in mind where the magnetic energy density is small compared with the gas pressure. We assume length scales for the flow field $L_\perp$ perpendicular to the field and $L_\parallel$ parallel to the field. For a small displacement ξ in the direction of the flow the change in the field is, in the first (curl free) case, of the order $\delta B = B_0\xi_\perp/L_\perp$. The change in magnetic energy is $\delta E_m = 1/2(\xi_\perp/L_\perp)^2 B_0^2/8\pi$. The kinetic energy is $E_k = 1/2\rho(v_\perp^2 + v_\parallel^2)$. The magnetic field starts interfering with a displacement of the amplitude chosen when $\delta E_m > E_k$. Since the flow is incompressible, $v_\parallel/v_\perp = L_\parallel/L_\perp$, so that this inequality yields

$$v_A > v_\perp(L_\perp^2 + L_\parallel^2)^{1/2}/\xi_\perp$$

For overturning motions (but not for, say, waves), the displacement ξ in one correlation time is of the order $(L_\perp, L_\parallel)$, so that magnetic stabilisation occurs for

$$v_A > v, \tag{1}$$

where $v = (v_\perp^2 + v_\parallel^2)^{1/2}$, for motions that are curl free in a plane perpendicular to the field. For curly (divergence free) motions in this plane, on the other hand, $\delta E_m = 1/2(\xi_\perp/L_\parallel)^2 B^2/8/\pi$, and the same argument yields stabilization for

$$v_A > vL_\parallel/L_\perp. \tag{2}$$

If the motions are roughly isotropic, both kinds of motions are affected as soon as the Alfvén speed exceeds the flow speed, but (2) shows that there is a special class of motions that is less affected. These are divergence free in a plane perpendicular to the field and their length scale along the field is larger than perpendicular to the field. That is, they are field aligned rolls.

We conclude that interference with hydrodynamic processes occurs when the Alfvén speed exceeds the flow speed. Its first effect is that it tries to force the motions into a pattern of field-aligned rolls. This behavior is well known also from a number of hydromagnetic stability problems (Chandrasekhar, 1961).

In order to find out more about the possible field strengths in the core, we now consider some theoretical considerations.

The magnetic torque

The spindown rate $\dot{\Omega}$ of the core is related to the radial and azimuthal field components at its surface by

$$\langle \sin^2\theta \; B_\phi B_r \rangle = -\frac{4\pi}{3}k^2\rho\,\dot{\Omega}\,r^2, \tag{3}$$

where the gyration constant k^2 is related to the moment of inertia I by $I = k^2Mr^2$, and the brackets denote average over the core's surface. If the cause of the corotation of the core with the solar surface is indeed a magnetic field, the present spindown rate, $\dot{\Omega} = 3\;10^{-6}\;\mathrm{s}^{-1}/\;3\;10^9$ yr, implies

$$\langle B_\phi B_r \rangle^{1/2} \sim 1\mathrm{G}. \tag{4}$$

The absolute value of a field satisfying (4) has a minimum when B_ϕ and B_r are the same and of the order 1G. Is a field of this magnitude plausible? To find out, we need to consider the processes that have evolved the field from its initial state when the sun was formed to the present. In the following I discuss the effects thought to be relevant in order of increasing complication. For earlier reviews, see Mestel (1984), Mestel and Moss, (1977), Mestel and Weiss (1987).

Time scales

The magnetic evolution is determined by two time scales, the Alfvén travel time

$$t_A \sim R/v_A = \frac{R}{B}\,(4\pi\rho)^{1/2} \tag{5}$$

where R is the radius of the core, ρ a typical value of the density. With $R = 5\ 10^{10}$ cm, $\rho = 1$ g cm^{-3}, $t_A \sim 1.5\ 10^{10}/B$ s. The magnetic diffusion time scale is

$$t_d = L^2/\eta,$$

where η is the diffusivity, and L the length scale on which the field changes. For $L = R$, $\eta \sim 10^4$ cm^{2}s^{-1}, this is of the order of the age of the present sun. Since several other processes work on shorter time scales, I leave out magnetic diffusion first. Processes that create small length scales in the magnetic field, such that diffusion can not be ignored also exist; these will be considered further on.

If the magnetic field is not in mechanical equilibrium, or if its equilibrium is unstable, the field will change into a different configuration on the Alfvén time scale. This may be compared with some of the Sun's other internal time scales. The dynamical time scale:

$$t_o \sim (GM_\odot/R^3)^{-1/2} \sim 2\ 10^3\text{s}, \tag{6}$$

the rotation period:

$$t_r = 2\pi/\Omega \sim 2\ 10^6\text{s}, \tag{7}$$

the spindown time scale:

$$t_s \sim t_\odot \sim 10^{17}\text{s}.$$

The magnetic time scale (5) is of the order of the dynamical time scale for a field of the order 10^8 G. Such a field has a significant effect on the structure of the sun, so that it would be visible in the frequencies of the solar p-modes. Upper limits on such structure effects are known (Dziembowski and Goode, 1990); they exclude a field as high as this. A field of 3 10^5 G has an Alfvén time scale of the order of the rotation period. It would cause changes in the p-mode frequencies of the order of the rotational splitting. Magnetic effects of this magnitude may be observable in the near future (Dziembowski and Goode, 1990). A field that is just strong enough to keep the core corotating with the surface under the action of the solar wind torque, must satisfy (3). If the radial and azimuthal field components are of the same order of magnitude, this implies that

$$t_A \sim (\Omega^{-1} t_s)^{1/2}, \tag{8}$$

which happens at $B \sim 1$G. Finally, a field for which the Alfvén time scale is equal to the spindown time is of the order $B \sim 1\mu$G.

Evolution of a weak field by differential rotation

We first consider the case when no external torque is present; this tells us how a field might reduce an initial nonuniform rotation. As an example, suppose that the initial field is axisymmetric about the rotation axis, and purely poloidal (azimuthal component zero). The rotation is nonuniform, with a characteristic difference in rotation rate $\Delta\Omega$ across the core. Since the field lines are frozen in (the magnetic diffusion time being long), the differential rotation winds the poloidal field up in the azimuthal direction, creating a B_ϕ component

$$B_\phi \sim B_p t \Delta\Omega \tag{9}$$

We now make an essential simplification by assuming that the poloidal component of the field does not change in time (for deviations from this see 'instabilities' below). Under this (somewhat restrictive) assumption, the field evolution can be calculated in some detail.

Initially, the internal torques are zero because B_ϕ is zero. As B_ϕ increases, a torque builds up that resists further winding up of the field. B_ϕ and this torque (proportional to $B_\phi B_p$) both increase linearly with $t\Delta\Omega$, the number of 'differential' rotations. The torque therefore causes the motion to oscillate harmonically, periodically reversing the differential rotation on a time scale

$$t_{Ap} = R/v_{Ap} = \frac{R}{B_p}\,(4\pi\rho)^{1/2}, \tag{10}$$

the Alfvén time based on the poloidal component of the field. This period is of the order of the age of the sun for an initial field of 1μG. Since the initial field is likely to be much higher than this (see below), the differential rotation will execute many oscillations. In this case, a strong *damping* of the oscillations by 'phase mixing' takes place.

Damping of magnetic oscillations

Since the initial state assumed above was axisymmetric, each of the poloidal field lines corresponds to an axisymmetric magnetic surface (a continuous surface everywhere parallel to the field). The oscillation can be seen as an Alfvén wave traveling on this surface. The energy propagates parallel to this surface, so that there is no coupling between adjacent magnetic surfaces. Since the Alfén speed varies through the star, the oscillation period defined loosely by (10) is in fact different for each magnetic surface. Adjacent surfaces therefore get out of phase after several periods, and the magnetic field becomes inhomogeneous across the surfaces on a length scale that decreases inversely proportional with time. Magnetic diffusion, even if negligible on the scale of the core, will therefore become dominant after a finite amount of time. Damping of waves by this process is called 'phase mixing' or 'Landau damping'. For references and an application to the solar corona see Heyvaerts and Priest (1983).

Assume that t_{Ap} varies by a factor of order unity between magnetic surfaces, across the core. Then two surfaces separated by a distance L will be out of phase by half a period after a number n of periods of the order $n = R/L$. Resistive damping becomes effective in one period when the Alfvén time scale is of the order of the diffusive time on the scale L:

$$L^2/\eta \sim R/v_{Ap}.$$

With this value for L, we find that diffusion damps the oscillation after $n = (R\, v_{Ap}/\eta)^{1/2}$ periods, corresponding to a damping time of

$$t_{damping} = (R^3/\eta\, v_{Ap})^{1/2}. \tag{11}$$

For a poloidal field of 1 G this is only 10^7 yrs, and one would expect any differential rotation in the core to be damped on this time scale. Calculations of the damping process have been made by Roxburgh (1987).

Nonaxisymmetric initial field

We write the field as a superposition of two parts, $\mathbf{B} = \mathbf{B_0} + \mathbf{B_1}$ such that $\mathbf{B_0}$ is the $m = 0$ component in a Fourier expansion in the azimuthal angle ϕ, and $\mathbf{B_1}$ the rest. Hence $\mathbf{B_0}$ is axisymmetric, and the azimuthal average of $\mathbf{B_1}$ is zero. The dynamic response of this field to the differential rotation is too complicated, but if the problem is considered in the *kinematic* limit, that is for fields that are weak enough that the Lorentz force can be ignored, the evolution of the field becomes tractable. Following Rädler (1983), and Zeldovich *et al.* (1985) we note that because of the kinematic assumption, the induction equation is linear in $\mathbf{B}$, so that it is sufficient to consider the evolution of $\mathbf{B_0}$ and $\mathbf{B_1}$ separately. Since $\mathbf{B_0}$ is axisymmetric, it does not change in time. The gradient in Ω however winds $\mathbf{B_1}$ up such that it develops ever decreasing length scales in the direction of $\nabla\Omega$. As in the phase mixing problem considered before, finite resistivity will become effective on these small length scales after a finite time, and thereafter the $\mathbf{B_1}$ decays rapidly to its azimuthal average, which is zero. Thus, this kinematic process smoothes the initial field to an axisymmetric field. The time scale for this to happen is the same as (10) but with $1/\Delta\Omega$ replacing the Alfvén travel time t_A:

$$t_{smooth} = (R^2/\eta\, \Omega)^{1/2}. \tag{12}$$

For $\delta\Omega$ of the order of the present rotation rate, this time scale is only 10^4 yr. If the initial field is weak enough such that the Lorentz force is still negligible after t_{smooth}, the evolution of the entire field is simple: first there is a phase in which the field is smoothened to an axisymmetric state, then the differential rotation acting on this will follow the evolution sketched in the previous section. This second phase removes the azimuthal component of $\mathbf{B_0}$, leaving only its poloidal part. Hence what is left after both processes is a *uniform rotation and an axisymmetric poloidal field.*

Closed field lines

In the simple picture sketched above, we have implicitly assumed that all field lines cross the surface of the core. If this is not the case, i.e. if there are closed field lines inside the core, then the final state, under the assumptions made is not necessarily one of uniform rotation. In this case (cf. Roxburgh, 1963; Mestel and Weiss, 1987) the rotation in the final state only has to be constant on each of the closed field lines (Ferraro's law of isorotation). Only the volume traced by the field lines that cross the surface would then corotate with the surface. If all field lines cross the surface and this surface is assumed to rotate uniformly, the end result would be uniform rotation throughout the core.

The spindown problem

In the preceding sections, we have ignored an external torque on the star, and considered only the evolution following an initial nonuniform distribution of the rotation rate. Adding a stellar wind torque does not change the problem fundamentally if the spindown time scale is much longer than the time scales $t_{damping}$ and t_{smooth}. The processes described are then shortlived transients set up by the initial conditions, the final state is a slowly evolving configuration of nearly uniform rotation (with the caveat outlined in the previous paragraph), with an axisymmetric field for which the torque balance relation (3) holds.

Strength of the initial field

The simple picture above required, among other things, that the initial field be weak enough, so that the initial evolution of the field can be calculated kinematically. How likely is this to be the case? To find out, the process of star formation itself must be considered. Present observational evidence (e.g. Shu *et al.* 1987) shows that in the final stage of star formation, the protostar is surrounded by a disk from which it accretes and continues to grow. A remnant of the magnetic field of the protostellar cloud is still embedded in this disk, and is carried in with the gas. This compresses the field so that close to the star it must become dynamically important and resists being further amplification by inward advection. In this state there is a balance between radially outward drift of the field under the influence of its Lorentz forces and inward advection. For a given accretion rate, the field strength at the protostellar surface can be calculated from this balance. For accretion at a rate of $10^{-5}\ M_\odot\ \mathrm{yr}^{-1}$ (believed to be realistic for the main accretion phase) it turns out to be of the order 10^3 G. We conclude that rather strong initial fields, much above the one Gauss level, may well be realistic.

Instabilities

Another assumption made above is that only the Lorentz force in the azimuthal direction needs to be considered. This component then determines the evolution of the internal rotation. In reality, the components in the meridional plane may be equally important. These would give rise to motions *the meridional plane* in addition to the pure rotation, and would change the field in much more complicated ways. No attempt has been made so far to describe such a situation realistically. Even if the initial state is one in which the field is perfectly in equilibrium so that there are no flows in r and θ directions, this equilibrium need not be stable.

Large classes of field configurations in a star are known to be dynamically unstable. For example, all purely poloidal fields are dynamically unstable (so that they would change significantly on the Alfvén travel time) in the absence of rotation (Wright, 1973; Tayler, 1980; Flowers and Ruderman, 1977; Goossens *et al.* 1986). Also all purely *toroidal* (azimuthal) fields are dynamically unstable without rotation (Goossens and Tayler, 1980; Van Assche *et al.* 1982). Purely toroidal fields are probably also unstable in the presence of uniform rotation (Pitts and Tayler, 1985). Since the Alfvén travel time is so short for the kind of fields (1G and up) considered here, the issue of stability is in fact crucial both for the state of rotation and for the nature of the field configuration to be expected in the core. The effects of a dynamic instability would be so severe that none of the scenario sketched above would be relevant.

At present, not enough is known about the stability problem to make much progress. From the stability results so far, all indicating instability, one might guess that instability is the rule, and that the interaction of the field with rotation takes place in the presence of a continuous dynamic instability. This view is biased however, as pointed out by Moffat at this meeting. The unstable fields found are all fields without magnetic helicity (the integral of $\mathbf{A} \cdot \mathbf{B}$ over the volume, where $\mathbf{A}$ is the vector potential of $\mathbf{B}$) because they have either a poloidal or a toroidal component but not both. It was pointed out already by Mestel and Moss (1977) that a dynamically stable field configuration most likely has both toroidal and poloidal components of roughly equal strength. In terms of the helicity, this can be formulated as follows. In the absence of magnetic diffusion, the magnetic helicity of a field confined in some closed volume is constant in time (Woltjer, 1958). Thus if the initial equilibrium has a finite helicity, its conservation implies that among all the configurations that can be reached from it by ideal displacements (without magnetic diffusion or reconnection) there is a non vanishing field with a finite minimum energy, which is therefore also a stable configuration. For more on the magnetic helicity constraint see Berger, 1988.

Such stable, linked poloidal- toroidal configurations have not been found yet because of the limitations of analytical methods in constructing magnetic equilibria. But knowing that they exist from the helicity argument, it may be worth trying to construct them by numerical means. In any case it would be psychologically very important to have at least one example of a stable magnetic equilibrium in an idealized (say, polytropic) star.

Evolution with instabilities

In this section we follow the consequences of a drastic assumption, namely that the field is permanently dynamically unstable, evolving on an Alfvén time scale (see also Spruit, 1987). For example, because it might turn out that stable equilibria would be rare so that most stars miss them in their evolution, or because over the life of the sun magnetic diffusion is still significant, so that Woltjer's theorem would not apply.

We ignore the geometrical properties of the field and model the instability by a simple exponential decay at a rate equal to the time scale of dynamical instability. These assumptions most likely overestimate the rate at which the field decays due to instabilities, so that the field strengths found by this line of reasoning will be lower limits.

In the absence of rotation the time scale of dynamical instability is just the Alfvén travel time (5). If the rotation period is less than this time scale, Coriolis forces can not be neglected, and the growth time of the instability is lengthened by a factor of the order Ωt_A (see Pitts and Tayler, 1985 for examples). This is because the dominant Coriolis force requires the flow developing in the instability to be mainly perpendicular to the direction of the magnetic force, so that much less energy is extracted from the driving force than in the absence of rotation. With this estimate of the growth time, a field that survives in the core until time t must have an instability time scale of the order t or longer:

$$t_A(\Omega t_A) > t \qquad \text{or} \qquad t_A > (\Omega t)^{1/2}.$$

Since at any time t the spindown time scale of the star is of the same order as its age, this estimate is equivalent to (8). The field strength corresponding to this, as discussed above, is also just the value required to keep the core corotating with the surface, and the torque balance equation (3) shows that the radial and azimutal field strengths are of similar magnitude for such a field. For the present sun, the permanent instability scenario therefore yields a field strength on the order of 1G, with the poloidal and azimuthal components of similar strength. This statement is rather coarse, since it neglects almost all the geometrical properties of the field, which are likely to play a significant role. Also, it is a lower limit to the field strength to be expected if stable field configurations exist into which the evolution of the configuration can get trapped.

Interference with mixing: an example

As an example of the kind of turbulence to be expected, we consider Zahn's circulation driven turbulence (see Zahn, this volume). In this scheme, twodimensional turbulence on equipotential (horizontal) surfaces has a tail of threedimensional motions at small scales. For mixing, the most important motions are those whose length scale is just at the transition between threedimensional and twodimensional flow. At this length, the velocity amplitude is

$$u_3 = (\epsilon_t/\Omega)^{1/2},$$

where ϵ_t is the energy input into the turbulence per gram per second, and Ω the rotation rate. For driving by the Eddington- Sweet- Vogt circulation this is:

$$u_3 = \Omega \; R \; (\tau_{KH}\Omega)^{-1/2} \frac{\Omega}{N} \; (\frac{R}{H})^{1/2}, \tag{13}$$

where N and H are the buoyancy frequency and the pressure scale height in the core, and τ_{KH} its Kelvin–Helmholtz time scale. Thus for the magnetic field to start interfering with this process we must have

$$B > B_c = (4\pi\rho)^{1/2}\Omega\ R(\tau_{KH}\Omega)^{1/2}\frac{\Omega}{N}\ (\frac{R}{H})^{1/2}.$$

For the present sun, $\rho \sim 1$, $\Omega \sim 3\ 10^{-6}$, $\tau_{KH} \sim 3\ 10^{14}$, $N \sim 10^{-3}$, $R \sim 5\ 10^{10}$, $H \sim 5\ 10^{9}$ (cgs), so that $B_c \sim 0.1$ G.

Bottom line

We conclude that under pessimistic assumptions about the stability of magnetic fields, the field strength in the present core is expected to be at least 1G. Such field strengths are sufficient to start interfering with the hydrodynamic mixing processes proposed. We may therefore be forced to consider the Lithium mixing problem as an intrinsically magnetohydrodynamic problem.

References

Berger, M. A. 1988, *Astron. Astrophys.* **201**, 355.

Brown, T. M. Christensen-Dalsgaard, J., Dziembowski, W. A., Goode, P., Gough, D. O. and Morrow, C.A. 1989, *Astrophys. J.* **343**, 526.

Christensen-Dalsgaard, J., Gough, D. O. and Thompson, M. J. 1989, preprint.

Chandrasekhar, S. 1961, Hydrodynamic and Hydromagnetic Stability, Clarendon Press (Dover Edition: 1981).

Drobyshevskii, E. M. and Yuferev, V. S. 1974, *J. Fluid Mech.* **65**, 33.

Dziembowski, W. and Goode, P. R. 1990, in Inside the Sun, G. Berthomieu and M. Cribier Eds., Kluwer Dordrecht, p. 341.

Flowers, E. and Ruderman, M. A. 1977, *Astrophys. J.* **215**, 302.

Galloway, D. J. and Weiss, N. O. 1981, *Astrophys. J.* **243**, 945.

Goossens, M. and Tayler, R. J. 1980, *Mon. Not. Roy. Astron. Soc.* **193**, 833.

Goossens, M., Poedts, S., and Hermans, D. 1986, *Solar Phys.* **102**, 51.

Heyvaerts, J. and Priest, R. E. 1983, *Astron. Astrophys.* **117**, 220.

Law, W. Y., Knobloch, E. and Spruit, H. C. 1984, in Observational tests of stellar evolution theory, A. Maeder and A. Renzini Eds., Reidel, Dordrecht, p. 523.

Mestel, L. 1961, *Mon. Not. Roy. Astron. Soc.* **122**, 473.

Mestel, L. and Moss, D. L. 1977, *Mon. Not. Roy. Astron. Soc.* **178**, 27.

Mestel, L. 1984, *Astron. Nachr.* **305**, 301.

Mestel, L. and Weiss, N. O. 1987, *Mon. Not. Roy. Astron. Soc.* **226**, 123.

Pitts, E. and Tayler, R. J. 1985, *Mon. Not. Roy. Astron. Soc.* **216**, 139.

Parker, E.N. 1963, *Astrophys. J.* **138**, 552.

Rädler, K. 1983, preprint.

Roxburgh, I. W. 1963, *Mon. Not. Roy. Astron. Soc.* **126**, 67.

Roxburgh, I. W. 1987, work in progress.

Schatzman, E. 1962, *Ann. Astrophys.* **25**, 18.

Schatzman, E. 1977, *Astron. Astrophys.* **56**, 211.
Shu, F. H., Adams, F. C. and Lizano, S. 1987, *Ann. Rev. Astron. Astrophys.* **25**, 23.
Spruit, H. C. 1987, in The Internal Solar Angular Velocity, eds. B.R. Durney and S. Sofia, p. 185 (Reidel, Dordrecht).
Tassoul, J.-L. and Tassoul, M. 1989, *Astron. Astrophys.* **213**, 397.
Tayler, R. J. 1980, *Mon. Not. Roy. Astron. Soc.* **191**, 151.
Vainshtein, S. I., Zeldovich and Ya. B. 1972, *Usp. Fiz. Nauk* **106**, 43 (Sov. Phys. Usp. **15**, 106 (1972)).
Van Assche, W., Tayler, R. J. and Goossens, M. 1982, *Astron. Astrophys.* **109**, 106.
Woltjer, L. 1958, *Proc. Nat. Acad. Sci. USA* **44**, 489.
Wright, G. A. E. 1973, *Mon. Not. Roy. Astron. Soc.* **162**, 339.
Zeldovich, Ya. B., Ruzmaikin, A. A. and Sokoloff, D. D. 1985, Magnetic fields in astrophysics, Gordon and Breach, New York.

FAST DYNAMO ACTION-A CRITICAL PROBLEM FOR SOLAR MAGNETISM

H. K. Moffatt
Department of Applied Mathematics and Theoretical Physics
University of Cambridge, UK

Abstract. Dynamo action is the process by which kinetic energy is systematically converted to magnetic energy in electrically conducting fluids. This process may occur on a resistive time-scale (i.e. with a growth rate that tends to zero as the magnetic resistivity of the medium tends to zero) in which case the dynamo is described as "slow"; or it may occur on the convective time-scale, in which case the dynamo is described as "fast". The conventional description of the solar dynamo (an $\alpha\omega$ dynamo in the terminology of mean-field electrodynamics) places it in the "fast" category, because the usual expressions for the regeneration parameter α and the turbulent resistivity β are both independent of molecular diffusivity η under the conditions obtained deep down in the solar convection zone. Current attempts to construct fast dynamos, and to understand their detailed structure, are therefore of critical importance in the context of solar magnetism.

It has been shown (Moffatt and Proctor, 1985) that in a fast dynamo, the length-scale of the magnetic field is necessarily nearly everywhere a factor of $R_m^{-1/2}$ smaller than the length-scale of the underlying convection process. In the case of the sun, with a magnetic Reynolds number R_m of order 10^4, this implies that the scale of the magnetic field that is generated may be as small as 10km, a result that is not incompatible with observation.

The dynamo process being turbulent, it remains of the greatest importance to have a reliable theory for determination of the parameter α and β in the limit of large magnetic Reynolds number. A renormalisation group procedure involving repeated averaging over successively increasing length scales (as described by Moffatt, 1983) seems to offer the best hope. In this theory, molecular diffusivity is important in the early stages of averaging, but becomes less important as the process is repeated, and the ultimate values of α and β are quite independent of η.

These problems are discussed, and some of the outstanding difficulties in the understanding of fast dynamos are identified.

References

Moffatt, H. K. 1983, *Rep. Prog. Phys.* **46**, 621.
Moffatt, H. K. and Proctor, M.R.E. 1985, *J. Fluid Mech.* **154**, 493.

INTERPRETING THE OBSERVATIONS

OBSERVATIONAL CONSTRAINTS ON THE TURBULENT DIFFUSION COEFFICIENT IN LATE TYPE STARS

Annie Baglin
Département d'Astrophysique Stellaire et Galactique, URA CNRS 335
Observatoire de Meudon
92195 MEUDON CEDEX FRANCE

and

Pierre Morel
Département G.D. Cassini,
Observatoire de la Cote d'Azur
BP139 06003 NICE CEDEX FRANCE

It has been already shown in several papers that a mild turbulent mixing is needed to explain the trend of the variation with effective temperature of the Lithium abundances in young clusters (see for instance Baglin *et al.* 1985, Schatzman 1981). Though a general qualitative agreement is possible, many uncertainties remain due essentially to our uncomplete knowledge of the different parameters entering the modelisation (Baglin and Lebreton 1990). However, in the case of the G and K stars of the Hyades observed by Cayrel *et al.* (1984), we will show here that some constraints can be derived for the turbulent diffusion coefficient.

1. Modelisation of the Hyades main sequence

To study the lithium depletion in the Hyades, models of zero age main sequence (ZAMS) are sufficient for two main reasons. First, the age dependence of the lithium abundance, easily seen when comparing clusters of different ages implies that the influence of the pre-main sequence is very small. Second, the Hyades are quite young (6 to 7 10^8 years), so that the evolution during the main sequence around one solar mass has not changed the structure of the model.

The observational main sequence is well defined, as well as the rotation curve (Fig 1). However, the distance remains uncertain within 15 to 20%, i.e. between 40 and 47 parsecs. Spectroscopic determinations of the abundances lead to a metal content somewhat higher than the sun [Fe/H]$\approx$0.14, which means Z$\approx$0.03 instead of 0.02. Some authors have also proposed a low value of the helium content Y, i.e. a large quantity of hydrogen X (Strömgren *et al.*, 1982).

In the HR diagram, the fitting of the models remains controversial as several parameters are not known and others are measured with a low accuracy. Fig. 2 shows the influence of these parameters on the position of the ZAMS, using some improved physical data: the last version of the Los Alamos Opacity Library, complemented by a low temperature part including molecular effects for the opacities. The net of nuclear reactions, and the treatment of the degeneracy of the equation of state are quite coarse, but we have checked that in this mass range they introduce differences in the models much smaller than the observational uncertainties. Assuming that the calibration of the mixing length obtained for the Sun is valid also for the Hyades (this means that the convection properties are somehow independent of age and chemical composition), the (Z=0.02 X=0.68) set of models represents quite correctly the observations. The Z=0.03 sequence looks too luminous and is shifted in mass by approximately 0.1 M⊙ ; this effect could then be compensated by a very small value of the distance, or an extremely large value of the mixing length parameter α, or a larger value of Y, hypothesis which all seem very unprobable. The presence of lithium in the coolest objects at $\log T_e$=3.71 as observed by Cayrel *et al.* (1984) can be used to favor some combination of these parameters. To survive at the age of the Hyades, superficial lithium has to be protected from the layer where it burns, i.e. around 2.7 10^6 K. The convective zone has then to be shallow enough to disconnect the surface material from this burning region. The structure of several models which occupy the low temperature end of the sequence is given in Table 1, where α has been kept constant; T_{ZC} is the temperature at the bottom of the convective zone.

Table 1. Characteristics of the different ZAMS models at the low temperature end

model	mass	Z	X	log L	log T_e	$T_{ZC}(10^6 K)$
1	1.0	0.03	0.68	-.238	3.725	2.6
2	0.9	0.02	0.68	-.300	3.724	2.34
3	0.9	0.03	0.64	-.349	3.715	2.74
4	1.0	0.02	0.73	-.217	3.726	2.24
5	1.1	0.03	0.73	-.168	3.768	2.62

It is well known (see i.e. Cayrel 1981) that looking only at the global parameters L and T_e, an increase of the metal content can be compensated by an increase of the mass (models 1 and 2, or 4 and 5); an increase of the metallicity can also be compensated by a decrease of the hydrogen content X, i.e. an increase of the helium content Y. But, as seen on Table 1, the temperature at the bottom of the convective zone decreases when X increases and Z decreases. Models with a high value of Z, as indicated by the observations, would have a high temperature T_{CZ} at the bottom of the convective zone, and lithium would be unable to survive. If our modelisation is correct (in particular the opacities) such a composition should then be excluded. A compensation of the effect of Z can be obtained by an increase of X, i.e. a decrease of Y, as proposed by Strömgren *et al.* (1982). But this chemical composition leads to very luminous models as compared to the observations even taking into account the uncertainty on the distance. Then the models which correspond to the best fit of the luminosity and of the extension of the convective zone, seem to be in conflict with the observed chemical composition.

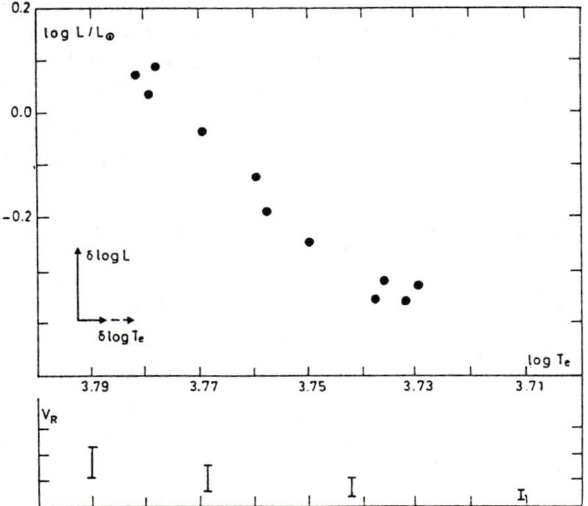

Figure 1. HR diagram of the Hyades adapted from Cayrel (1981) and rotation curve from Stauffer (1987). Average uncertainties on L and Te are indicated in the lower left corner.

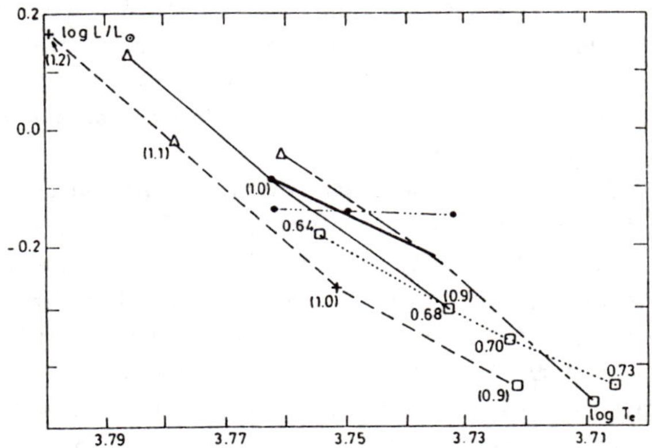

Figure 2. Theoretical Zero Age Main Sequences for different values of the parameters for solar type stars. Masses are given in parenthesis; models with the same mass, the same Z and different values of X are joined by a dotted line; models with Z=0.02 and X=0.68 by a continuous line, models with Z=0.03 and X=0.68 by a dashed-dotted line , models with diferent values of α by the horizontal line. The bold line represents the shift from Z=0.02 to Z=0.03, all other parameters being the same. The dotted line represents the ZAMS of Lebreton et al. (1987)

2. Constraints on the diffusion coefficient

The modelisation as well as the method of resolving the diffusion equation has been described in Baglin *et al.* (1985, 1987). In the following we have chosen a set of parameters (X=0.70, Z=0.02, α=2.2, molecular opacities included) very close to the the classical solar ones as determined by Lebreton and Maeder (1987), because they correspond to the best fit. Table 2 indicates the global parameters of the zero-age sequence of models and the corresponding range of observed rotationnal velocities.

Table 2. Characteristics of the models of G and K Hyades stars

model	mass	log L	log T_e	V_{rot}(km/s)
A	.9	-.432	3.711	3 - 6
B	1.0	-.284	3.742	4 - 10
C	1.1	-.013	3.769	5 - 15
D	1.2	.170	3.789	10 - 22
E	1.3	.340	3.810	20 - 30
F	1.35	.490	3.826	50

The dependence of the diffusion coefficient on the physical parameters is still very uncertain, though several important steps have been achieved (see i.e. Zahn 1987, Vauclair 1989, Pinsonneault *et al.* 1989). Instead of trying to fit predictions of the surface abundance using a given formula for the diffusion coefficient including different parameters (radius, density, rotationnal velocity....) to the observed Lithium abundances, we ask here a somewhat different question: *what is the average value of this coefficient needed to reproduce the observed abundances for the G and K stars of the Hyades, how does it vary with mass?*.

Fig 3 shows this mean value as a function of mass. An almost constant value seems to be needed, in contradiction with the values obtained with Zahn's formula.

Fig. 4 confirms this result: the predicted abundances of lithium as function of the effective temperature using Zahn's coefficient are too high; the diffusion coefficient is too small at low temperature. A constant diffusion coefficient over the entire range of temperature is not sufficient at the lower end either; an extension of the convective region corresponding to some overshooting is needed. The size P of the region of penetration of the convective motions into the underlying radiative zone is mimicked by an overshooting parameter O_v, such that P=$O_v * H_p$ where H_p is the pressure scale height. Fig. 5 represents the mean value of the diffusion coefficient as a function of the depth of the mixing region. This result can be interpreted as follows: the influence of the penetration at the bottom of the convective zone increases as it becomes shallower.

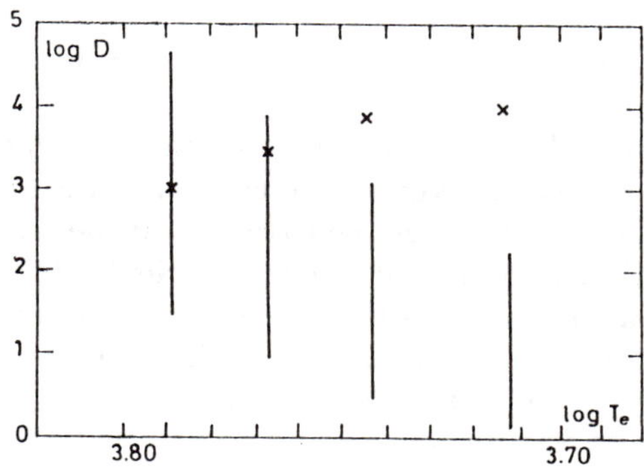

Figure 3. Mean value of the diffusion coefficient necessary to reproduce the observed Lithium abundances as a function of effective temperature(crosses), compared to the interval covered by the diffusion coefficient computed according to Zahn's value (vertical bars); models identification are given in parenthesis.

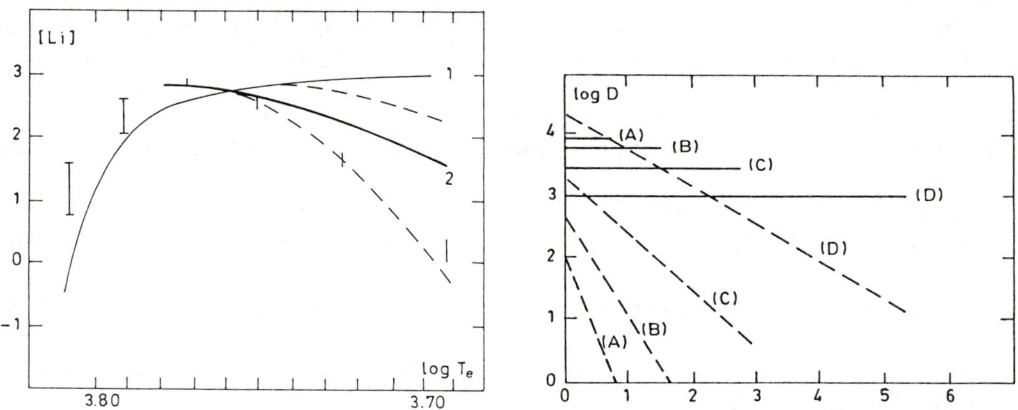

Figure 4. Predicted Lithium abundances as a function of effective temperature using a constant diffusion coefficient equal to 2500 (2) and Zahn's one (1) compared to the observed values (vertical bars). Two values of the penetrative convection parameter Ov, have been tested: Ov=0 full lines, and Ov=0.7 dotted lines.

Figure 5. Mean value of the diffusion coefficient (full line) compared to the Zahn values (dotted line) as a function of the depth of the mixing region measured in Hp units for differnet models defined in Table 2.

3. Conclusions

The information contained in the temperature and age dependance of the lithium abundances leads to severe constraints on the description of the physics of the outerlayers in solar type stars. At the present stage, constraints can be put on both the chemical composition and the behavior of the diffusion coefficient. However, some caution is needed as many uncertainties remain in the modelisation. More precise determinations of the distances (HIPPARCOS) as well as more reliable computations of the opacities in the 10^4 to $2\ 10^6$ K temperature range will help.

Acknowledgements : We would like to thank G. and R. Cayrel, E. Schatzman for fruitful discussions, L. Leon and F. Tran-Minh for making available in a computable form the Los Alamos opacity tables.

References

Baglin, A., Lebreton, Y. 1990, in Inside the Sun, IAU colloquium **121**, G. Berthomieu and M. Cribier eds., Kluwer Academic Publishers, p. 437.

Baglin, A., Morel, P.J., Schatzman, E. 1985, *Astron. Astrophys.* **149**, 309.

Baglin, A., Morel P.J. 1987, in the Impact of Very High S/N Spectroscopy on Stellar Physics, IAU Symp. No. 132, eds G. Cayrel and M. Spite, Kluwer Academic Publishers 1988, p. 279.

Cayrel de Strobel, G. 1981, in Stars clusters, J.E. Hesser ed., p 91.

Cayrel, R., Cayrel de Strobel, G., Campbell, B., Däppen, W. 1984, *Astrophys. J.* **283**, 205.

Lebreton, Y., Maeder, A. 1987, *Astron. Astrophys.* **175**, 99.

Pinsonneault, M.H., Kawaler, S.D., Sofia, S., Demarque, P. 1989, *Astrophys. J.* **338**, 424.

Schatzman E. 1981, in Turbulent diffusion and the solar neutrino problem, CERN 81-11.

Stauffer, J.R. 1987, in Formation and evolution of low mass stars, eds. A.K.Dupree and MTVT Logos, p. 331.

Strömgren, B., Olsen, E.H., Gustafsson, B. 1982, *Publ. Astr. Soc. Pac.*, **94**, 5.

Vauclair, S. 1989, *Astrophys. J.* **335**, 971.

Zahn, J.P. 1987, in The internal angular velocity of the Sun, B.R. Durney and S. Sofia eds., Reidel, Dordrecht, p. 201.

PROBING STELLAR INTERIORS

Sylvie Vauclair
Observatoire Midi-Pyrénées,
Toulouse, France

What can the chemical composition of stellar atmospheres tell us about transport processes and macroscopic motions inside the stars?

Two different classes of mechanisms may lead to modifications of the chemical composition of stellar atmospheres during the star's life.

The first one is the tandem: "nuclear reactions and macroscopic motions": the chemical composition is modified inside the star due to nuclear reactions, but these modifications do not appear at the surface unless either some transport process occurs in the stellar outer layers, or the star is "pealed off" by mass loss. The first case occurs, for example, in Red Giants where the products of the CNO cycles may be seen at the surface, and in cool Main-Sequence and Giant stars in which lithium is depleted. The second case occurs in Wolf-Rayet stars and some White Dwarfs.

The second type of mechanisms is only possible when the stellar gas is specially stable against macroscopic motions: then the chemical elements may selectively diffuse due to non-equilibrium processes. These processes are induced by the pressure and temperature gradients, and the radiative tranfer. This is, for example, the reason of the abundance anomalies observed in the so-called Chemically Peculiar Stars. In the absence of macroscopic motions, the abundance variations due to selective diffusion could be computed without any arbitrary parameter. However macroscopic motions do occur in stars. They have to be introduced in the computations, with some parameters, the values of which are not known from hydrodynamical studies. Comparisons between the abundances computed in this frame and the observed abundances can thus lead to interesting constraints on these parameters. Considering that the abundances of all the elements present in a star have to be consistently matched, we may hope that the number of constraints exceed the number of free parameters. This is why the study of the chemical composition of stellar atmospheres is powerful for our understanding of macroscopic motions and transport processes in stars.

I will not give a review of this very large subject. I will here only focus on two examples of the connexion between atmospheric abundances and transport processes in main sequence stars:

-rapidly oscillating Ap stars, in relation with their chemical composition and the observation of helium-rich stars,

-lithium depleted main sequence stars, and specially the problem of the "lithium gap" observed in young clusters.

1. Rapidly oscillating Ap stars.

The rapidly oscillating Ap stars discovered by Kurtz (1982) lie among the coolest magnetic chemically peculiar stars. They oscillate with periods between 4 and 15 minutes and the amplitude of the light variations vary, at least in some of them, in phase with the magnetic field. These observations lead Kurtz to suggest an explanation of these stars in terms of an oblique pulsator model, namely non radial pulsations aligned with the magnetic axis, this magnetic axis itself being inclined with respect to the rotation axis. Since 1982, many observations have been pursued by Kurtz and co-workers, leading to more and more precise constraints on the periods and amplitudes of the pulsation modes: see for example the review by Kurtz (1985).

Several models have been suggested to explain these pulsations : Dolez and Gough (1982) proposed an excitation mechanism related to the abundance anomalies which are observed in these stars, and which are presumably due to element segregation. They discussed the fact that if helium was overabundant at the magnetic poles, it could trigger the pulsations in the direction of the magnetic axis, and lead to unstable modes as observed. They fell across an apparent inconsistency, as element segregation leads to helium depletion at the magnetic poles while helium accumulation should be necessary for triggering the pulsations. Shibahashi (1983) suggested magnetic overstability at the poles as an excitation mechanism for the oscillations. Mathys (1985) proposed that, contrary to Kurtz's oblique pulsator model, the oscillations could be aligned with the rotation axis instead of the magnetic axis, and that the amplitude modulations could be due to abundance spots on the star, which would block part of the flux. None of these ideas lead to completely satisfactory explanations of the observations.

Dolez and Gough (1982) suggestion has been revisited by Vauclair and Dolez (1989) and Dolez, Gough and Vauclair (1990), with the added hypothesis of a stellar wind at the magnetic poles. Helium gravitational settling in a stellar wind was invoked by Vauclair (1975) to account for H_e-rich stars, which are observed at the hot end of the magnetic star sequence. These 20000K to 25000K stars exhibit a helium enrichment by a factor 2 to 3 which cannot be explained by diffusion alone, the upward radiative force on helium being smaller than the downward gravitational force (see Montmerle and Michaud, 1976). Vauclair (1975) showed that an overall mass loss flux of order $10^{-12}M_{\odot}.yr^{-1}$ could convect helium up to the observed layers and leave it there due to the larger effect of gravitational settling when helium is neutral: then the collisions are about 50 times less effective than when it is once ionized, and helium is no more pulled out by the hydrogen flux. Evidences of the occurence of outflowing winds controlled by large magnetic fields in chemically peculiar stars were obtained later by Barker *et al* (1982) for the H_e-rich star HD 184927 and by Brown, Shore and Sonnebrown (1985) for the CP2 star HD 21699 (B6V). IUE spectra present a blueshifted component of CIV lines in both of these stars. The line profiles vary with time, in phase with the magnetic field, as expected (see also the dicussion on the magnetically controlled winds by Casinelli and Lamers, 1987). Using the model developped by Castor *et al* (1981), Michaud *et al* (1987) deduce from these observations a mass loss rate of at least $5x10^{-12}M_{\odot}.yr^{-1}$ at the magnetic poles. Such a mass loss rate is too large to allow for helium separation. However, as this rate is reduced from the magnetic pole to the magnetic equator due to the horizontal component of the magnetic lines, they suggest that helium enrichment occurs somewhere in between. The same process can apply for cooler magnetic stars as well.In case of helium gravitational settling in a mass loss flux, the helium outward motion may be reversed in the region where helium becomes neutral. However, for smaller effective temperatures, this region lies inside the star and not at the surface as in H_e-rich stars. The outer layers may appear H_e-poor, while helium accumulation occurs deeper in the star.

Figure 1 shows the "helium diffusion flux" in a 1.6 $M_\odot$ star. This "diffusion flux" is defined as the diffusion velocity multiplied by the density of the stellar gas. The decline by a factor 50 from the atmosphere to the base of the H_eI ionisation zone is due to the fact that neutral elements diffuse more rapidly than ionised ones as they suffer less collisions. When helium is once ionised, the diffusion flux varies like $T^{3/2}$ as expected from the diffusion theory for simple gases. A second minimum is obtained at the base of the H_eII ionisation zone, followed by a steady increase, again like $T^{3/2}$, once helium is completely ionised. This diffusion flux can be directly compared to the mass loss flux, which is represented as a horizontal line, in the plane-parallel approximation. It immediately appears that a mass loss flux of about $5\text{x}10^{-12}$ to $5\text{x}10^{-11}$g.cm^{-2}.s^{-1} will produce a helium enrichment in the layers where $\Delta M/M$ (ratio of the external mass to the stellar mass) is about 10^{-10}. Such a mass loss rate corresponds to 10^{-14} to $10^{-13} M_\odot.yr^{-1}$, which is quite realistic.

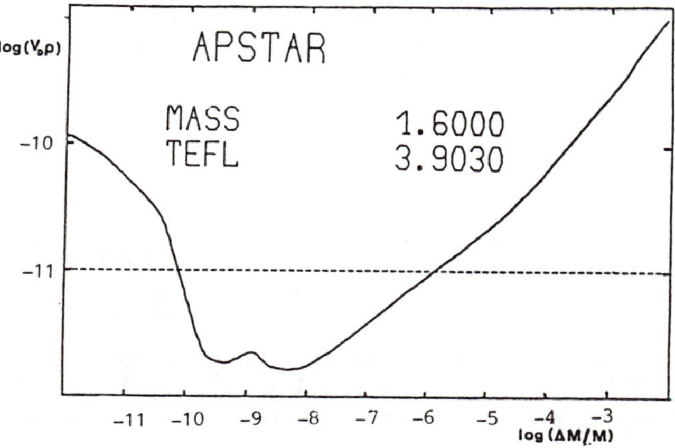

Figure 1. Helium diffusion flux versus depth in a 1.6.$M_\odot$ star. Δ M/M refers to the outer mass fraction, *i.e.* the ratio of the mass of the outer zone to the total mass of the star. $V_{D\rho}$ represents the helium diffusion velocity multiplied by the gas density (downward flux). The dashed line represents a mass loss flux of 2 x 10^{-14} $M_\odot.yr^{-1}$ or 10^{-11}g.cm^{-2}.s^{-1} (upward flux). Such a wind leads to a helium accumulation around Δ M/M = 10^{-10} and depletion above that layer.

In case of no turbulence helium would accumulate at this place, creating a sharp helium peak in the first ionisation zone, which would be of no use for triggering the pulsations, as the κ-mechanism is basically effective in the second helium ionisation zone. However such a peak would not be stable as it would create a strong density inversion. The general problem of the thermohaline instability induced by helium accumulation is not trivial. First order computations show that helium should be mixed down in a very short time scale (of the order of one year). However helium accumulation by a factor 2 or 3 is indeed observed in the external layers of H_e-rich stars, so that it *must* be possible in spite of the presumptions of instabilities (the fact that H_e-rich stars lie on the hydrogen main-sequence excludes that they be helium-rich throughout). The stabilisation of the H_e rich layer may be due to the presence of the magnetic field. Some partial mixing must however occur, otherwise the helium accumulation would be orders of magnitude larger than observed. We parametrize turbulence below the helium peak by a turbulent diffusion coefficient which is constant below the helium maximum abundance and vanish above it to modelize the destabilizing effect of the positive μ-gradient and the stabilizing effect of the negative one.

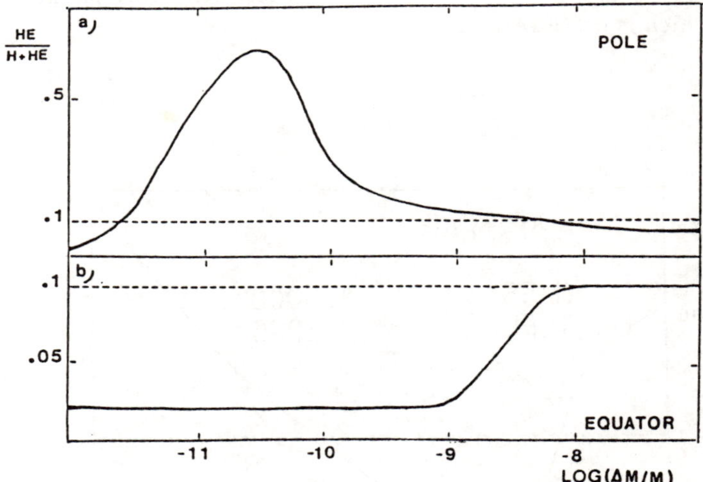

Figure 2. Examples of helium abundance profiles obtained at the magnetic poles and the magnetic equator of a 1.6 $M_\odot$ star, with a mass loss flux of 2 x $10^{-14} M_\odot.yr^{-1}$ and a turbulent diffusion coefficient below the peak of $10^5 cm^2.s^{-1}$. *(after Dolez and Vauclair, 1989).*

Figure 2a shows the helium accumulation which can be expected in the helium ionisation zones of the polar regions with a turbulent diffusion coefficient of 10^5 below the helium peak. Overabundances by factors 2 or 3 can be obtained in this case as well as in helium rich stars. Meanwhile helium falls below the convection zone in the equatorial regions (figure 2b). The large helium peak obtained in the first helium ionisation zone does not trigger the pulsations. However the small increase (by a factor of about 1.2) which remains in the second helium ionisation is enough to destabilize the star in this region (figure 3). It is thus possible to account for non spherically symetric pulsations in Ap stars due to the non spherically symetric chemical composition. More computations are needed however to do quantitative comparisons between observed and computed modes. In any case the observational evidences of helium rich layers in stars give constraints on the efficiency of thermohaline convection in a vertical magnetic field. It would be interesting to study this problem in more details from a hydrodynamical viewpoint.

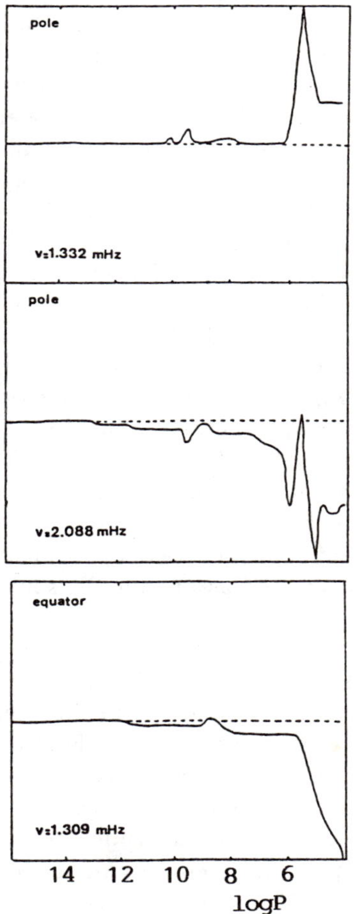

Figure 3. Examples of energy curves inside a 1.6 $M_\odot$ star for three different modes, obtained with the profiles shown on figure 2: an unstable mode at the magnetic pole (top), a stable mode at the magnetic pole (middle), and a stable mode at the equator (bottom). *(computations by Dolez and Vauclair, 1989).*

2. Lithium depleted main-sequence stars.

The "lithium gap" observed in the Hyades and other galactic clusters by Ann Boesgaard and her collaborators, and confirmed by several authors (Boesgaard and Tripicco, 1986; see also the review of the observations by Charbonneau and Michaud, 1988), gives a challenge to theoreticians. At the present time, three different explanations have been given for this gap.

Michaud (1986) proposed to explain this gap by selective lithium diffusion. In G type stars, the convection zone is too deep for gravitationnal settling to take place: the density at the bottom of the convection zone is so large that the diffusion time scale exceeds the age of the star. Increasing the effective temperature leads to a decrease of the convection zone, and consequently to a decrease of the diffusion time scale. In F stars it becomes smaller than the stellar age, leading to a lithium abundance decrease as observed. When the convection zone is shallow enough, the radiative acceleration on lithium becomes important as lithium is in the hydrogenic form of L_iIII (while it is a bare nucleus, L_{rmi}IV, deeper in the star. This radiative acceleration may prevent lithium settling for hotter F stars. However it would lead to significant lithium overabundance in early F stars, which is not observed. Michaud (1986) suggests that mass loss can then prevent this lithium accumulation. Another difficuly with this model is that it needs a stable gas below the convection zone, while turbulence is needed to account for the lithium decrease in G stars (see, for example, Baglin *et al*, 1985; Cayrel *et al*, 1984).

Other explanations of the lithium gap, in terms of mixing and nuclear destruction, have been suggested by Vauclair (1987, 1988) and Charbonneau and Michaud (1988). These explanations followed a remark by Boesgaard (private communication), who pointed out that the lithium gap occurs just in the region (in effective temperature) where the stellar rotational velocities increase. Vauclair (1988) treated the rotational mixing in the frame of Zahn (1987) theory of instabities induced by meridional circulation. Assuming that the stars in the Hyades have always rotated with their present rotational velocity, the lithium decrease on the red side of the gap is well reproduced in the right time scale (figure 4). Charbonneau and Michaud (1988) obtained similar results within the frame of the Tassoul and Tassoul (1982) theory of meridional circulation.

The assumption of constant angular rotation for a given spectral type is somewhat crude and has to be discussed. It is currently believed, from observations of rotation velocities of T Tauri and main-sequence stars and from theories of angular momentum loss, that stars first accelerate during the T Tauri phase due to their contraction, and then decelerate due to magnetically driven mass loss (see Schatzman, 1990). The outer convection zone is probably first decelerated in a short time scale (about 10^7 years), and then angular momentum is transported inside the star so that it finally rotates approximately as a solid body. This seems to occur in a time scale shorter than the age of the Hyades (Kawaler, 1988). Computations of lithium destruction in pre main sequence stars may be found in D'Antona and Mazzitelli (1984) and Proffitt *et al* (1989)). Without "extra mixing" (other than convection), the effect is negligible. Computations of lithium destruction induced by rotation in these stars are not yet available. Pinsonneault *et al* (1989) have modelized the mixing induced by transport

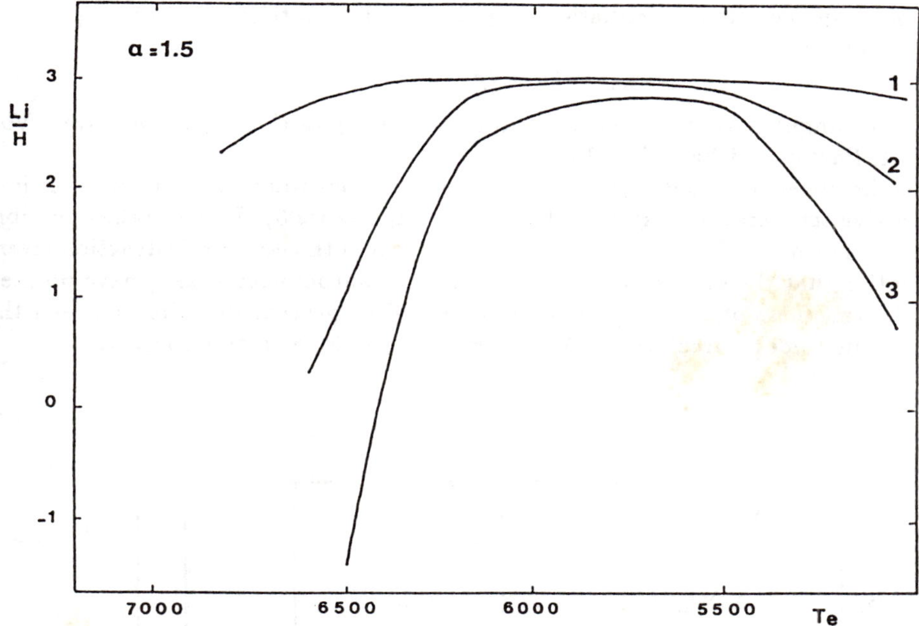

Figure 4. Theoretical lithium abundance curves in main sequence population I stars computed in the frame of Zahn's theory, with a semi-implicit numerical diffusion code. The convection zones are computed within the mixing length approximation, with α, ratio of the mixing length to the pressure scale height, of 1.5. Curves are drawn for 8 x 10^7 yrs, 8 x 10^8 yrs and 4 x 10^9 yrs. *(after Charbonnel, Vauclair and Zahn, 1990).*

of angular momentum in solar type stars and its incidence on the lithium abundance. If the transport of angular momentum was due to plain turbulence, no lithium would be left in solar type stars. Assuming that the transport of matter is proportional to the transport of angular momentum, they show that, for lithium to be depleted by one hundred as observed in the sun, the efficiency of matter mixing must be 50 times smaller than the efficiency of angular momentum transport. The assumption of constant rotation velocity in stars of a given spectral type requires a negligible matter mixing during the phase of angular momentum transport. This would be the case, for example, if this transport was essentially induced by a magnetic field inside the star.

The two approaches for the explanation of the lithium gap by nuclear destruction and mixing differ for what concerns the blue side, and the fact that the lithium abundance is normal in hot F stars. Charbonneau and Michaud (1988) have computed the radiative acceleration on lithium in F stars and argue that it can then compete with the effect of the meridional circulation: in spite of this circulation, lithium do not fall down and so it may remain unchanged in the outer layers. Vauclair (1988) have noticed that two separate zones of meridional circulation may develop in rotating stars, the separation occuring at the layer for which:

$$\Omega^2/4\pi G\rho = 1$$

where Ω is the angular rotation velocity and ρ the density in the considered layer (Von Zeipel, 1924; Pavlov and Yakovlev, 1977).

This "quiet zone" lies inside the convection zone for cool stars, and gets out of it just for the effective temperature corresponding to the gap (figure 5). It seems thus possible that matter is not mixed between the convection zone and the nuclear destruction layers for stars hotter that the gap. However Michaud and Charbonneau (1990) have pointed out that the thickness of this quiet zone is too small to prevent diffusion between the two loops of meridional circulation. More work is being done on this subject.

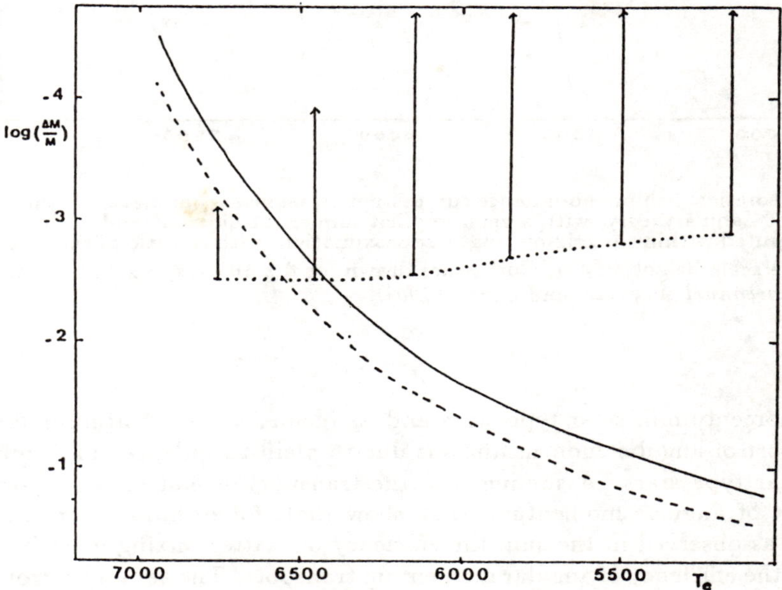

Figure 5. Position of the "quiet zone" between the two loops of meridional circulation in population I main sequence stars. The ordinate represents the ratio of the external mass to the mass of the star. The solid line is the bottom of the outer convection zone for a = 1.9. The dashed line is the bottom of the mixed zone if overshooting is added over one pressure scale height. The dotted line represents the position of the "quiet zone" if the star rotates with a velocity of 100 $Km.s^{-1}$. The ends of the arrows show the position of this zone when the stars rotates at the present Hyades velocities : 50 $Km.s^{-1}$ for the first one on the left (1.35 $M_\odot$), 30 $Km.s^{-1}$ for the second one (1.3 $M_\odot$) and 10 $Km.s^{-1}$ for the other ones (in this case the end of the arrows are off scale). *(after Charbonnel, Vauclair and Zahn, 1990).*

Another problem remains, which has been pointed out by Michaud (private communication). If either the Tassoul and Tassoul meridional circulation theory or the Zahn rotationally induced turbulence theory are extrapolated for A stars, it seems that the macroscopic motions are too large to allow gravitationaly and radiatively driven diffusion. Constraints on the turbulent diffusion coefficients which could allow selective diffusion have been given by Michaud *et al* (1976) and Vauclair *et al* (1978). The most constraining elements are those which appear as underabundant: they may be depleted only in a small fraction of the star, where they are in the noble gas configuration, and turbulence can rapidly transform an underabundace in an overabundance. The observations of calcium and scandium in Am stars (Cayrel de Strobel, 1990) are extremely important in that respect.

In summary, a lot of work is still needed both on the observational and on the theoretical sides to reach a consistent view of the stellar hydrodynamics. The present situation looks like a big jig-saw puzzle whose pieces are not yet in place, but for which the astrophysical community has made much progress during the recent years.

References.

Baglin, A., Morel, P.J., Schatzman, E. 1985, *Astron. Astrophys.* **149**, 309.

Barker, P.K., Brown, D.N., Bolton, C.T., Landstreet, J.D. 1982 in Advances in Ultraviolet Astronomy: Four years of IUE research (NASA CP-2238), p. 589.

Boesgaard, A.M., Tripicco, M.J. 1986, *Astrophys. J.* **303**, 724.

Brown, D.N., Shore, S.N., Sonneborn, G. 1985, *Astron. J.*, **90**, 1354.

Casinelli, J.P., Lamers, H.J.G.L.M. 1987, in Exploring the Universe with the IUE Satellite, Y. Kondo Ed., Reidel, Dordrecht, p.139.

Castor, J.I., Lust, J.H., Seaton, M.J. 1981, *Mon. Not. Roy. Astron. Soc.*, **194**, 547.

Cayrel, R., Cayrel de Strobel, G., Campbell, B., Däppen, W. 1984, *Astrophys. J.* **283**, 205.

Cayrel de Strobel , G., 1990, this conference.

Charbonneau,P., Michaud, G. 1988, *Astrophys. J.* **334**, 746.

Charbonnel, C., Vauclair, S., Zahn, J.P. 1990, in preparation.

D'Antona, F., Mazzitelli, I. 1984, *Astron. Astrophys.*, **138**, 431.

Dolez, N., Gough, D.O. 1982, in Pulsations in Classical and Cataclysmic Variable Stars, J.P.Cox and C.J. Hansen Eds., JILA, Boulder, p. 248.

Dolez, N., Gough, D.O., Vauclair, S. 1990, in preparation.

Kawaler, S.D. 1988, *Astrophys. J.* **333**, 236.

Kurtz, D.W. 1982, *Mon. Not. Roy. Astron. Soc.*, **200**, 807.

Kurtz, D.W. 1985, in Seismology of the Sun and Distant Stars, D.O. Gough Ed., Reidel, p. 441.

Mathys, G. 1985, *Astron. Astrophys.* **151**, 315.

Michaud, G. 1986, *Astrophys. J.* **302**, 650.

Michaud, G., Charland, Y., Vauclair, S., Vauclair, G. 1976, *Astrophys. J.* **210**, 447.

Michaud, G., Dupuis, J., Fontaine, G., Montmerle,T. 1987, *Astrophys. J.* **322**, 302.

Michaud, G., Charbonneau, P. 1990, preprint.

Montmerle, T., Michaud,G. 1976, *Astrophys. J. Suppl.*, **31**, 489.

Pavlov, G.G., Yakovlev, D.G. 1977, *Astr. Zh.*, **55**, 1043.
Pinsonneault, M.H., Kawaler, S.D., Sofia, S., Demarque, P. 1989, *Astrophys. J.*, **338**, 424.
Proffitt, C.R., Michaud, G. 1989, *Astrophys. J.*, **346**, 976..
Schatzman, E. 1990, this conference.
Shibahashi, H. 1983, *Astrophys. J.*, **275**, L5.
Tassoul, J.L., Tassoul, M. 1982, *Astrophys. J. Suppl.*, **49**, 317.
Vauclair, S. 1975, *Astron. Astrophys.* **45**, 233.
Vauclair, S. 1987, in Atmospheric Diagnostics of Stellar Evolution: Chemical Peculiarities, Mass Loss, and Explosion, ed. Nomoto, Springer Verlag.
Vauclair, S. 1988, *Astrophys. J.* **335**, 971.
Vauclair, S., Dolez, N. 1989, in Progress of Seismology of the Sun and Stars, Shibahashi Ed., Springer-Verlag.
Vauclair, S., Vauclair, G., Schatzman, E., Michaud, G. 1978, *Astrophys. J.* **223**, 567.
Von Zeipel, H., 1924, *Mon. Not. Roy. Astron. Soc.*, **84**, 665.
Zahn, J.P. 1987, in the Internal Solar Angular Velocity, B. R. Durney and S. Sofia Eds., Reidel, Dordrecht, p. 201.

CONCLUSION

The meeting ended with an informal session, held outdoor in the resplendent Sun, during which the participants drew the conclusions of what they had heard and understood. Most of those conclusions have been incorporated in the individual contributions of these proceedings, and it suffices to summarize the most salient remarks.

As could be expected, the need for more complete, more homogeneous and more precise observations was stressed. Of the rotation rates, which can now be inferred from spectral variations, when there are some irregularities at the surface of the stars. Of the accretions disks surrounding the young stars, trying to determine when they disappear in the course of the evolution. Of the lithium abundance (both isotopes) in stars of all ages, including binaries; comparing similar stars, such as the solar analogues. Of the beryllium abundance, whenever that determination is possible. Of the lithium abundance in the interstellar medium, in the Magellanic Clouds.

On the theoretical side, it was felt that much effort should go into securing a better description of the dynamics and the hydrodynamics in convective envelopes. How can turbulence be characterized there? Where is the magnetic field generated? What picture can we draw of the region which is just below a convection zone? What is the precise role of rotation, of differential rotation? How is the loss of angular momentum linked with the intensity of the magnetic field and with the rotation rate?

We decided to gather again with Evry Schatzman, in a not too distant future, hopefully to rejoice over the progress made meanwhile in answering all these questions.

Jean-Paul Zahn

Lecture Notes in Mathematics

Vol. 1236: Stochastic Partial Differential Equations and Applications. Proceedings, 1985. Edited by G. Da Prato and L. Tubaro. V, 257 pages. 1987.

Vol. 1237: Rational Approximation and its Applications in Mathematics and Physics. Proceedings, 1985. Edited by J. Gilewicz, M. Pindor and W. Siemaszko. XII, 350 pages. 1987.

Vol. 1250: Stochastic Processes – Mathematics and Physics II. Proceedings 1985. Edited by S. Albeverio, Ph. Blanchard and L. Streit. VI, 359 pages. 1987.

Vol. 1251: Differential Geometric Methods in Mathematical Physics. Proceedings, 1985. Edited by P.L. García and A. Pérez-Rendón. VII, 300 pages. 1987.

Vol. 1255: Differential Geometry and Differential Equations. Proceedings, 1985. Edited by C. Gu, M. Berger and R.L. Bryant. XII, 243 pages. 1987.

Vol. 1256: Pseudo-Differential Operators. Proceedings, 1986. Edited by H.O. Cordes, B. Gramsch and H. Widom. X, 479 pages. 1987.

Vol. 1258: J. Weidmann, Spectral Theory of Ordinary Differential Operators. VI, 303 pages. 1987.

Vol. 1260: N.H. Pavel, Nonlinear Evolution Operators and Semigroups. VI, 285 pages. 1987.

Vol. 1263: V.L. Hansen (Ed.), Differential Geometry. Proceedings, 1985. XI, 288 pages. 1987.

Vol. 1265: W. Van Assche, Asymptotics for Orthogonal Polynomials. VI, 201 pages. 1987.

Vol. 1267: J. Lindenstrauss, V.D. Milman (Eds.), Geometrical Aspects of Functional Analysis. Seminar. VII, 212 pages. 1987.

Vol. 1269: M. Shiota, Nash Manifolds. VI, 223 pages. 1987.

Vol. 1270: C. Carasso, P.-A. Raviart, D. Serre (Eds.), Nonlinear Hyperbolic Problems. Proceedings, 1986. XV, 341 pages. 1987.

Vol. 1272: M.S. Livšic, L.L. Waksman, Commuting Nonselfadjoint Operators in Hilbert Space. III, 115 pages. 1987.

Vol. 1273: G.-M. Greuel, G. Trautmann (Eds.), Singularities, Representation of Algebras, and Vector Bundles. Proceedings, 1985. XIV, 383 pages. 1987.

Vol. 1275: C.A. Berenstein (Ed.), Complex Analysis I. Proceedings, 1985–86. XV, 331 pages. 1987.

Vol. 1276: C.A. Berenstein (Ed.), Complex Analysis II. Proceedings, 1985–86. IX, 320 pages. 1987.

Vol. 1277: C.A. Berenstein (Ed.), Complex Analysis III. Proceedings, 1985–86. X, 350 pages. 1987.

Vol. 1283: S. Mardešić, J. Segal (Eds.), Geometric Topology and Shape Theory. Proceedings, 1986. V, 261 pages. 1987.

Vol. 1285: I.W. Knowles, Y. Saitō (Eds.), Differential Equations and Mathematical Physics. Proceedings, 1986. XVI, 499 pages. 1987.

Vol. 1287: E.B. Saff (Ed.), Approximation Theory, Tampa. Proceedings, 1985–1986. V, 228 pages. 1987.

Vol. 1288: Yu. L. Rodin, Generalized Analytic Functions on Riemann Surfaces. V, 128 pages, 1987.

Vol. 1294: M. Queffélec, Substitution Dynamical Systems – Spectral Analysis. XIII, 240 pages. 1987.

Vol. 1299: S. Watanabe, Yu.V. Prokhorov (Eds.), Probability Theory and Mathematical Statistics. Proceedings, 1986. VIII, 589 pages. 1988.

Vol. 1300: G.B. Seligman, Constructions of Lie Algebras and their Modules. VI, 190 pages. 1988.

Vol. 1302: M. Cwikel, J. Peetre, Y. Sagher, H. Wallin (Eds.), Function Spaces and Applications. Proceedings, 1986. VI, 445 pages. 1988.

Vol. 1303: L. Accardi, W. von Waldenfels (Eds.), Quantum Probability and Applications III. Proceedings, 1987. VI, 373 pages. 1988.

Lecture Notes in Physics

Vol. 343: M. Month, S. Turner (Eds.), Frontiers of Particle Beams; Observation, Diagnosis and Correction. Proceedings, 1988. IX, 509 pages. 1989.

Vol. 344: M. Rascle, D. Serre, M. Slemrod (Eds.), PDEs and Continuum Models of Phase Transitions. Proceedings, 1988. VII, 229 pages. 1989.

Vol. 345: H. Holden, A. Jensen (Eds.), Schrödinger Operators. Proceedings, 1988. V, 458 pages. 1989.

Vol. 346: D. Lüst, S. Theisen, Lectures on String Theory. VII, 346 pages. 1989.

Vol. 347: Ph. Blanchard, J.-P. Dias, J. Stubbe (Eds.), New Methods and Results in Non-linear Field Equations. Proceedings. 1987. VII, 133 pages. 1989.

Vol. 348: A. Bohm, M. Gadella, Dirac Kets, Gamow Vectors and Gel'fand Triplets. VII, 119 pages. 1989.

Vol. 349: Y.A. Kubyshin, J.M. Mourão, G. Rudolph, I.P. Volobujev, Dimensional Reduction of Gauge Theories, Spontaneous Compactification and Model Building. X, 80 pages. 1989.

Vol. 350: G. Tenorio-Tagle, M. Moles, J. Melnick (Eds.), Structure and Dynamics of the Interstellar Medium. Proceedings, 1989. XXI, 537 pages. 1989.

Vol. 351: A. Dervieux, B. Larrouturou (Eds.), Numerical Combustion. Proceedings, 1989. VII, 481 pages. 1989.

Vol. 352: K.B. Wolf (Ed.), Lie Methods in Optics II. Proceedings, 1988. XII, 197 pages. 1989.

Vol. 353: M. Barthes, J. Léon (Eds.), Nonlinear Coherent Structures. Proceedings, 1989. X, 280 pages. 1990.

Vol. 354: J. De Coninck, F. Dunlop (Eds.), Wetting Phenomena. Proceedings, 1988. VII, 112 pages. 1990.

Vol. 355: R. Lima, L. Streit, R. Vilela Mendes (Eds.), Dynamics and Stochastic Processes. Proceedings, 1988. V, 298 pages. 1990.

Vol. 356: R. Wehrse (Ed.), Accuracy of Element Abundances from Stellar Atmospheres. Proceedings, 1988. V, 109 pages. 1990.

Vol. 357: J.L. Lumley (Ed.), Whither Turbulence? Turbulence at the Crossroads. Proceedings, 1989. V, 535 pages. 1990.

Vol. 358: J. Kovalevsky, Astrométrie moderne. VI, 294 pages. 1990.

Vol. 359: K. Kirchgässner (Ed.), Problems Involving Change of Type. Proceedings, 1988. XII, 207 pages. 1990.

Vol. 360: Y. Mellier, B. Fort, G. Soucail (Eds.), Gravitational Lensing. Proceedings 1989. XV, 315 pages. 1990.

Vol. 361: P. Gaigg, W. Kummer, M. Schweda (Eds.), Physical and Nonstandard Ganges. Proceedings, 1989. IX, 310 pages. 1990.

Vol. 362: N.E. Kassim, K.W. Weiler (Eds.), Low Frequency Astrophysics from Space. Proceedings. XII, 280 pages. 1990.

Vol. 363: V. Ruždjak, E. Tandberg-Hanssen (Eds.), Dynamics of Quiescent Prominences. Proceedings, 1989. XI, 304 pages. 1990.

Vol. 364: T.T.S. Kuo, E. Osnes, Folded-Diagram Theory of the Effective Interaction in Nuclei, Atoms and Molecules. V, 175 pages. 1990.

Vol. 365: M. Schumacher, G. Tamas (Eds.), Perspectives of Photon Interactions with Hadrons and Nuclei. Proceedings. IX, 251 pages. 1990.

Vol. 366: M.-J. Goupil, J.-P. Zahn (Eds.), Rotation and Mixing in Stellar Interiors. Proceedings, 1989. XIII, 183 pages. 1990.